Succeed At
Numeracy
Tests
In A Week

Mac Bride saw his first book published in 1982, and since then he has written more than 120 books on various aspects of programming, computer applications, the Internet, languages for people buying houses overseas, green issues and other topics. As well as writing, he has edited and typeset books on subjects ranging from marketing through feng shui to allotment gardening. His recent books include *Great at my Job but Crap at Numbers* and the *IQ Workout Bullet Guide*. In this book he draws on that experience, and also on many years of teaching maths.

Teach Yourself®

Succeed At Numeracy Tests In A Week

Mac Bride

Typeset by Cenveo® Publisher Services.

Printed and bound in Great Britain by CPI Group (UK) Ltd, Croydon, CR0 4YY.

John Murray Learning policy is to use papers that are natural, renewable and recyclable products and made from wood grown in sustainable forests. The logging and manufacturing processes are expected to conform to the environmental regulations of the country of origin.

John Murray Learning
Carmelite House
50 Victoria Embankment
London
EC4Y 0DZ
www.hodder.co.uk

Contents

Introduction

When you apply for a job, you may well find that you have to sit psychometric tests. Many employers believe that these tests help them assess candidates' abilities and aptitudes, although their accuracy may be debatable. What they do measure, however, is a person's ability to do the tests. The good news is that you can improve this ability with training and practice. This is especially true of numeracy tests, where success depends as much on specific skills as on raw ability. This book aims to teach you those skills.

Numeracy tests can be designed to assess speed or power, or a combination of the two.

Speed tests can be based on computation or estimation, and they involve basic arithmetical functions, fractions, decimals and ratios. The questions are not normally difficult but have to be done against the clock. We'll tackle speed tests in the first three days of this week:

- **Sunday** – arithmetic, covering the four basic operations of addition, subtraction, multiplication and division, with an emphasis on estimates and quick ways to check answers. We'll also look at how you can increase accuracy by working backwards and rounding up and down.
- **Monday** – mixed sums, fractions and decimals. You will have done these things at school, but you may not have used them much since.
- **Tuesday** – percentages, ratios and averages. You'll learn how to calculate them and how to convert one type of measurement into another, as well as some useful shortcuts.

Power tests fall into two groups: numerical reasoning and data interpretation. In different ways these assess your ability to interpret information, analyse situations and solve problems logically. We spend the next three days working on these:

- **Wednesday** – 'real world' sums, or at least as 'real' as these things ever are. You'll see how to pick out the relevant numbers from the text and work out which type of calculation you need to solve the problem.
- **Thursday** – patterns in sequences and in layouts of numbers. Here the arithmetic is normally simple and, if you know the types of sequences and patterns that occur, you can spot them more easily.
- **Friday** – interpreting data from charts, graphs and tables: reading and making sense of them. You'll learn how to extract information from them quickly and accurately, so that you can make comparisons and do calculations. Good estimating skills usually help with these.

Every day you'll find exercises on each topic as it is covered and a longer test at the end of the day. The answers to these are at the end of each chapter.

- **Saturday** – test technique: how to make best use of your time and your strengths when sitting a numeracy test. There are mixed tests here to give you more practice on the things you have learned earlier in the week.

You can complete this book in a week if you spend a few hours at it every day, but you should achieve the same result if you take a month over it. While some of the tests should be done at speed because that is what they are testing, learning is best done at a pace that allows you to reflect on each new nugget of knowledge as you acquire it.

Enjoy, and good luck with your numerically tested job hunting!

Mac Bride

SUNDAY

Better, faster calculations

To score well in arithmetically based numeracy tests, you need speed and accuracy, but most of us find that the faster we work the more likely we are to make mistakes. Today you'll learn how to do sums faster without sacrificing accuracy.

The key to faster addition and subtraction lies in the complements of ten – the pairs of numbers that add up to ten. We all learned these at infant school, but many of us forget to use them or we fail to exploit them properly.

The key to faster multiplication and division is knowing your times tables but, before your heart sinks at the prospect of having to relearn them all, don't worry. You only need to learn three sets, and two of those are really easy.

There are two key techniques for spotting mistakes:

- estimating – recalculating the sum with nice round numbers to check the size of the answer
- end checks – recalculating the bits of the sum which produce the first and last digits.

Taken together, these will alert you to most mistakes.

Working in tens

The counting system is based on tens. This is because we have ten fingers (just as computers count in twos because they have two 'fingers' – electric switches can only be on or off). If you work with the tens, sums are simpler.

What a nice pair of numbers

Complements are good. (Not as in 'Nice hairdo!' That's a compliment.) The complements of ten are those pairs of numbers that add up to ten:

1 + 9	2 + 8	3 + 7	4 + 6	5 + 5
	6 + 4	7 + 3	8 + 2	9 + 1

You do know these. Just keep them in mind when you are doing addition and subtraction.

Addition

We're going to ignore single-digit addition; you can do those. We'll start with numbers of two or more digits. Always add the units, then the tens (and then the hundreds, etc., if there are any). Adding up is easier if one of the numbers ends in 0. Unfortunately, most of the time they won't.

The simplest sums are where the digits add up to less than ten, e.g.:

23 + 45   break the 45 into tens and units, and add the units
23 + 5 = 28   then add the tens
28 + 40 = 68.

Where the digits will add up to more than ten, break the units in the second number into two bits – one to complement the first, so that you have a multiple of ten. Like this:

27 + 48   split 48 into 3 (the complement of 7) and 45
27 + 3 + 45   add 3 to the first number (27), then add the tens
30 + 45 = 75.

An extension of the same trick can be used with bigger numbers:

264 + 387   break 7 into 6 (complement of 4) and 1,
264 + 6 = 270 + 381   turn to the tens and split 8 into 3 (complement of 7) and 5
270 + 30 = 300 + 351   and then add the hundreds to get 651.

Practice 1

1) 63 + 26   2) 36 + 57   3) 142 + 53   4) 55 + 36 + 113

You will find the answers at the end of the chapter.

Adding in columns

If the numbers have two or more digits and there are more than three numbers, then jot them down in a column, right aligned. So, if you have got this:

23 + 47 + 117 + 94

Set them out like this:

```
 23
 47
117
 94
```

Adding single digits down a column is easier than adding bigger numbers to each other. If you can do it more easily, you are more likely to do it right. You must make sure that the numbers are properly aligned, so that the units are in one column, the tens in one column, etc. Mix up the columns and you'll be in a real mess.

Practice 2

Set out in columns and add using the tens trick:

1) 25 + 68 + 63 + 92 + 47 + 12
2) 71 + 215 + 98 + 324 + 62
3) 34 + 145 + 232 + 571

Using a calculator

If a numerical test has anything more complicated than the sort of sums shown here, you will almost certainly be allowed to use a calculator. However, to use one properly, you need to be able to do simple arithmetic – you'll see why shortly – which is why this section is here.

Subtraction

Subtraction is addition in reverse, and it's very simple where the digits in the number you are taking away are smaller than those in the first number. For example:

48 – 25

We can do this in two simple steps. First the units:

(4)8 – 5 = (4)3

Here you are doing 8 – 5 but keeping the 40 in the back of your mind.

43 – 20 = 23

Here you are doing 4 – 2 (tens) with the 3 (units) on hold.

When there are bigger digits in the second number, you need a different approach: use the complements as described above.

46 – 28

First, focus on the units. Break the 8 into two parts: 6, which will take 46 down to 40, with 2 left over (8 – 6).

Take 2 from 40, to give you 38, then finish as before:

38 – 20 = 18.

Practice 3

1) 67 – 42   2) 39 – 28   3) 50 – 34   4) 72 – 56

The art of estimating

Good estimating skills can be a huge time saver when working with numbers. They will enable you quickly to spot wrong answers and identify right answers.

Estimates depend upon nice round numbers. A nice round number should only have one, or at most two, significant figures. A significant figure is one in the range 1 to 9. To make a number into a nice round number, round it up or down to the nearest 10, or 100, or 1,000 (or 10,000, etc.). If the second digit from the left is under 5, round down; if it is 5 or over, round up. Keep in mind which way you round the numbers as it can be useful later.

Here are some examples.

Original number	Nice round number	Keep in mind
26	30	quite a bit under 30
431	400	a fair bit over 400
61,038.23	60,000	not much over 60,000
3.1417	3	a tad over 3
8.75	10	quite close to 10

Notice those special technical terms 'quite a bit', 'a fair bit' and 'not much'. If the estimates are not much under or over the original numbers, then the end result won't be much different from the result you would get with the real figures. On the other hand, if the numbers are all 'quite a bit over' their nice round versions, the end accurate result will be a whole lot over the nice round answer.

To score well, you need to be able to calculate quickly and accurately and to recognize a right answer.

Checking your answers

If you use a calculator for adding up or subtracting a series of figures, how can you know that you haven't mis-keyed? The answer is to check the size and check the end. Here are some numbers to add up:

7,345 + 987 + 4,645 + 12 + 184 + 36 + 8,593

Your calculator gives you 21,802. Is this correct?

Check the size

First, we convert all those fiddly digits to nice round numbers and, as we do this, we will discard the smaller numbers altogether.

In 7,345, look at the first two digits: 73. Since 3 is less than 5 we can ignore it and set all the following digits to 0. In 987, the 8 is more than 5, so round up the 9 to 10. Set the other two digits to 0 and we have 1,000. Likewise, 4,645 becomes 5,000 and 8,593 becomes 9,000. As the answer is going to be in the thousands, we'll ignore 12, 36 and 184.

The sum now reads:

7,000 + 1,000 + 5,000 + 9,000

Now add those significant digits: 7 + 1 + 5 + 9 = 22
and tack on the 000s because these are thousands: 22,000.

We see that 22,000 is close to 21,802. This means that there are no major errors in the calculator value.

How much is a fair bit?

You can find the answer to this question, and discover how many tads make a smidgen, at http://www.crapatnumbers.net

Check the end

Next, we focus on the last digit in each number, and add these.

$5 + 7 + 5 + 2 + 4 + 6 + 3 = 32$

This tells us that the sum must end in 2. And our calculator result does. As the size is right, and the last digit is right, we can be confident that we have used the calculator correctly.

Practice 4

Use estimates and end checks to identify the correct answer to each of these sums:

1) $127 + 354 + 417 =$
   a) 889   b) 977   c) 898   d) 1,048
2) $1,028 + 42 + 584 + 221 =$
   a) 1,875   b) 975   c) 1,866   d) 2,253
3) $345 + 735 + 9,348 + 72 + 114 =$
   a) 41,088 b) 10,614 c) 10,532 d) 10,569
4) $3,486 - 234 =$
   a) 1,146   b) 3,248   c) 3,252   d) 282

Go forth and multiply

Don't panic! In numeracy tests you should only ever need to do simple multiplication – i.e. numbers of two or three digits by values up to ten – in your head. And you don't have to do the whole thing in your head – just the multiplying. Use a scrap of paper to record your workings and the result.

Does this mean that you have to learn (or relearn) your times tables? Yes and no. Here are some tricks that give you easy ways of multiplying by most of the numbers up to ten.

To multiply by	Trick
2	Double it.
3	No trick, but you should know your three times table already.
4	Double it, then double again.
5	Multiply by 10 and then halve the result: $7 \times 5 = 7 \times 10 = 70/2 = 35$.
6	Multiply by 3 and then double it – or the other way round.
7	Sorry, no trick! You have to learn these.
8	Double it, double it and double it again. (It's quicker than it sounds.)
9	Times 3, and times 3 again.
10	Add 0 to the end if it's an integer (whole number), or move the decimal point one place to the right.

Practice 5

1) 3×42
2) 13×4
3) 18×5
4) 25×6
5) 8×21
6) 15×9
7) 423×10
8) 24×7

Times tables: the only ones you need to learn									
1	2	3	4	5	6	7	8	9	10
2	4	6	8	10	12	14	16	18	20
3	6	9	12	15	18	21	24	27	30
4	8	12	16	20	24	28	32	36	40
5	10	15	20	25	30	35	40	45	50
6	12	18	24	30	36	42	48	54	60
7	14	21	28	35	42	49	56	63	70
8	16	24	32	40	48	56	64	72	80
9	18	27	36	45	54	63	72	81	90
10	20	30	40	50	60	70	80	90	100

TIP *You'll find handy print-out-and-keep times tables, in A4 and pocket/handbag sizes, at www.crapatnumbers.net*

Multiplying bigger numbers

In a test situation you are very unlikely to be asked to work out a complex multiplication sum by hand. You may be asked to work them out with a calculator, and you may be asked to pick the correct answer in a multi-choice. Estimates will help with both of these types of questions. For example, take:

476 × 293

Our calculator has given us 139,468. Is that right?

1 First check the size. We'll reduce these to nice round numbers: 500 × 300.
   a) Multiply the first digits: 5 × 3 = 15.
   b) Then add on a zero for every zero in the two original rounded numbers: 150,000.
   c) That's the same size as the calculator's answer. It has the same number of digits and there's not a huge difference between 150 and 139.
2 Second, multiply the last two digits: 6 × 3 = 18
3 The calculator answer should end with the same digit – and it does.

Practice 6

Use estimates and end checks to identify the correct answer to each of these sums:

1) 35 × 21
   a) 65   b) 735   c) 753   d) 420
2) 742 × 36
   a) 46,746   b) 212,412   c) 26,712   d) 27,621
3) 7,385 × 5,783
   a) 42,707,455   b) 27,952,225   c) 13,168   d) 27,952,252
4) 143 × 25 × 62
   a) 44,6166   b) 221,605   c) 888,150   d) 221,650

Division

Division is a combination of multiplication and subtraction. When you divide, you are working out how many times you have to multiply one number until it's as big as (or nearly as big as) the other. You are highly unlikely to be asked to work out a complex division in a numeracy test, but you may be asked to do simple division or to select the correct answer from several possible answers. A combination of using estimates and working backwards should bring you success with these.

Estimating division

There are three stages to this:

1 Find nice round numbers to replace the amount to be divided (the dividend), and the number you are dividing it by (the divisor). Thus, instead of 75 we use 80; instead of 12,345.89 we use 10,000.
$18,564 \div 537$ becomes $20,000 \div 500$

2 If there is a zero at the end of both numbers, knock it off them both, and keep doing this until one of them runs out of zeros.
$20,000 \div 500 = 2,000 \div 50 = 200 \div 5$

3 Lastly, run through multiples of the divisor (using your print-out-and-keep times table if necessary) until you get close to the amount to be divided. Don't sweat over this. If you've got a really large number divided by a smaller one, stop after the first couple of digits and fill out with zeros. That's your (near enough) answer.
$4 \times 5 = 20$, so $20 \div 5 = 4$ and $200 \div 5 = 40$

For example, say there are 137 reams of paper in the store cupboard. The firm normally gets through seven reams a day. How long will the stocks last? The sum is:

$137 \div 7$

Which rounds to

$140 \div 7$

A quick run through the 7 times table gets us to $2 \times 7 = 14$, so $20 \times 7 = 140$. As we had rounded up to 140, the answer is going to be a little under 20 days.

Working backwards

If you have an estimated answer, you can work back from this to find the correct answer. If you multiply the answer by the divisor (the one that you divided by), you should get back to the original number.

That last sum showed us that $137 \div 7$ is a bit under 20. Let's try 19×7. That gives us 133, which is 4 less than 137. The correct answer is 19 remainder 4.

Say you've got a report to print. Its word count is 9,607 and you can expect to print a little over 400 words per page. How many pages will it take? The sum is:

$9,607 \div 400$

This rounds to

$10,000 \div 400$

Knocking off the noughts gives you

$100 \div 4$

$2 \times 4 = 8$ and $3 \times 4 = 12$, so $20 \times 4 = 80$ and $30 \times 4 = 120$. Split the difference, and we see that $25 \times 4 = 100$.

There were a bit under 10,000 words, and a little over 400 fit on to a page, so we can expect the final page count to be more than a little under 25.

Practice 7

Use estimates and end checks to find the correct answer to each of these sums:

1) $288 \div 6 =$
   a) 43   b) 48   c) 33   d) 158

2) $5,697 \div 27 =$
   a) 211   b) 349   c) 71   d) 223

3) $1,152 \div 64 =$
   a) 18   b) 17   c) 204   d) 213

4) $2,244 \div 132 =$
   a) 23   b) 32   c) 17   d) 202

TIP *If you've forgotten the maths you (half-)learned at school and now find that you need it again, try this* Teach Yourself *book:* Great at my Job but Crap at Numbers, *by Mac Bride and Heidi Smith. Find out more at www. crapatnumbers.net*

Summary

Today's work was to remind you of some of the basic rules and techniques of arithmetic.

● Addition and subtraction are simpler if you use the tens complements.

● For anything other than the simplest addition and subtraction sums, you should jot them down on paper, making sure that the columns are properly aligned.

● You can check the size of answers by estimating them, using nice round numbers.

● Calculating with just the last digits is a second check on the accuracy of your answers.

● You should know your times tables, but you only really need to learn the twos, threes and sevens. All the rest can be worked out from these.

● In numerical tests you will be allowed to use a calculator if there are difficult multiplication or division sums, but you must check your answers with estimates and end checks.

MONDAY

TUESDAY

WEDNESDAY

THURSDAY

FRIDAY

SATURDAY

End-of-the-day test

[answers at the back]

This is a speed test. The sums are easy enough for everyone to be able to get them all correct, given enough time, but you don't have the time! You have 15 minutes in total. That's five minutes for each set. Go!

Calculate, in your head or on paper:

1) 27×6

2) $45 + 74 + 15 + 37 + 26$

3) $365 - 52$

4) 57×14

5) $164 + 86 + 274 + 7 + 21$

6) $2,636 - 1,529$

7) $84 \div 6$

8) $15 \times 12 \div 5$

9) $240 \div 60$

10) $7 \times 5 \times 4 \times 3$

Pick the correct answer:

11) $16 + 46 = 6 + ?$
    a) 55   c) 66
    b) 56   d) 67

12) $44 - ? = 15$
    a) 29   c) 28
    b) 26   d) 39

13) $210 \div 35 = ?$
    a) 5
    b) 6
    c) 7
    d) 8

14) $274 - 38 + 15 = ?$
    a) 271   c) 251
    b) 327   d) 297

15) $272 - ? = 145$
    a) 137   c) 217
    b) 417   d) 127

16) $57 + 24 = 35 + ?$
    a) 48   c) 42
    b) 46   d) 56

17) $74 - ? = 42$
    a) 22   c) 30
    b) 28   d) 32

18) $196 \div 7 = ?$
    a) 18   c) 28
    b) 27   d) 42

19) $327 - 18 + 103 = ?$
    a) 412   c) 242
    b) 359   d) 312

20) $63 \times 215 = ?$
    a) 1,345   c) 13,453
    b) 1,245   d) 13,545

True or false?

21) $13 \times 24 = 312$
    a) True   b) False

22) $76 \div 3 = 22$
    a) True   b) False

23) $828 + 452 = 1180$
    a) True   b) False

24) $37 + 3 - 15 - 7 + 21 - 14 = 25$
    a) True   b) False

SUNDAY

MONDAY

TUESDAY

WEDNESDAY

THURSDAY

FRIDAY

SATURDAY

25) 143 − 27 − 32 + 6 − 1 = 91
    a) True   b) False

26) 543 × 19 = 1,317
    a) True   b) False

27) 162 ÷ 6 = 27
    a) True   b) False

28) 694 + 725 = 1,429
    a) True   b) False

29) 5,382 − 2,478 = 2,904
    a) True   b) False

30) 729 × 305 = 22,345
    a) True   b) False

Answers to practice questions

Practice 1:

1) 89	3) 195
2) 93	4) 204

Practice 2:

1) 307	3) 982
2) 770	

Practice 3:

1) 25	3) 16
2) 11	4) 16

Practice 4:

1) c	3) b
2) a	4) c

Practice 5:

1) 126	5) 168
2) 52	6) 135
3) 90	7) 4,230
4) 150	8) 168

Practice 6:

1) b	3) a
2) c	4) d

Practice 7:

1) b. (Estimate: 300 ÷ 6 = 50 but a bit less than 300, so a bit less than 50. 48 is the right size. Work backwards to check the end: 8 × 6 = 48. The last digit matches.)

2) a. (Estimate: 6,000 ÷ 30 = 600 ÷ 3 = 200. Could be 211. Try end check: 1 × 7 = 7, so 211 is good.)

3) a. (Estimate: 1,200 ÷ 60 = 120 ÷ 6 = 20. Must be 18 or 17. Try end check for 18: 8 × 4 = 32. The last digit matches.)

4) c. (2,244 ÷ 132 rounds to 2,000 ÷ 100 = 20. Could be a or c. End check 23: 3 × 2 = 6, no good. End check 17: 7 × 2 = 4. The last digit matches.)

MONDAY

Mixed sums, fractions and decimals

Today we're doing some more arithmetic. First we will look at sums that involve different operations and learn the right order in which to tackle them. Next we will turn to fractions.

There are two ways of expressing parts of a number: fractions (e.g. $\frac{1}{2}$, $\frac{3}{4}$, $\frac{7}{10}$, $5\frac{7}{8}$) and decimal fractions (0.5, 0.625, 12.9). Fractions can be a useful way of describing simple divisions. How do you share three pizzas between five people? If you put three houses on half a hectare of land, how big is each plot?

The more complex the fractions, the harder they are to work with, which is where decimals come in (and you can't do sums with fractions on a calculator without converting them to decimals first). With decimals you are back to working with tens, which makes life much easier: as long as you keep good track of the decimal point, you shouldn't have a problem.

BODMAS, the rules of the sums game

If a sum has a mixture of addition, subtraction, multiplication and/or division operations, the way you work through it can change the outcome. To make sure everyone gets the same answer, there are rules that set the order in which to perform the operations. Computers follow these same rules. The mnemonic is BODMAS, which stands for these operations, and the order in which to carry them out:

1 **B**rackets
2 **O**f (to the power of)
3 **D**ivision and **M**ultiplication (if there are both, it doesn't matter which you do first)
4 **A**ddition and **S**ubtraction (which can also be done in either order).

$$\frac{47 \times (135 + 85.3)}{(387 - 215)}$$

Here's a simple example. Take this multi-operation sum:

$2 + 3 \times 4 - 1$

If you work through that left to right, you get:

$2 + 3 = 5 \times 4 = 20 - 1 = 19$

which is the wrong answer.

Here we go again, but following the BODMAS rules:

First do $3 \times 4 = 12$.

The sum is now $2 + 12 - 1 = 13$, which is the right answer.

If there are brackets in the sum, the operations inside them are performed first. Here are the same numbers again, but with some brackets:

$2 + 3 \times (4 - 1)$

First we do $(4 - 1)$. The sum is now $2 + 3 \times 3$.

Next we do the multiplication $(3 \times 3 = 9)$, then the addition: $2 + 9 = 11$, which is a different answer from last time.

Practice 1

Calculate the following, with BODMAS in mind and without using a calculator:

1) $2 \times 5 + 6 \div 4 - 3 =$
2) $2 + 5 \times 6 \div 4 - 3 =$
3) $(2 + 5) \times 6 \div (4 - 3) =$
4) $32 + 15 \times 3 - 9 =$
5) $(47 - 8) \div 3 =$

Calculators and BODMAS

Calculators don't know about the BODMAS rules, and don't normally have brackets. This means that sometimes you have to change the order in which you do things to get the right result.

Most pocket calculators process operations as they are entered. So, if you type in:

2 [+] 3 [×] 5 [=]

you will get 25, because it will work out $2 + 3 = 5$, then $5 \times 5 = 25$. According to BODMAS, the answer should be 17. To get $2 + (3 \times 5)$, start with the operation to be calculated first:

3 [×] 5 [+] 2 [=]

This produces $3 \times 5 = 15 + 2 = 17$.

Practice 2

Use a calculator to find the answers to these sums, and check them with estimates. If they are not round about the same size, have a careful think about how you are putting the sum into the calculator.

1) 175 + 26 × 12 + 378 =

2) 2,485 ÷ (93 − 22) =

3) 296 + (538 + 65 + 234) × 16 =

4) 600 − 4 × 72 − 23 × 3 =

5) 600 − 4 × (72 − 23) × 3 =

Visualizing fractions

To work happily with fractions, you need to have a good mental image of what they mean. When you see 2/5, in your mind's eye you should see something like 2 slices from a cake cut into 5 pieces, or for 7/10 you might see 7 bricks in a pile of 10. The image gives you the sense of proportion – which gives you the estimate that you need for checking. In this circle, for example, you should be thinking of 2/5 as a bit under half.

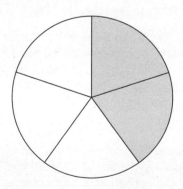

In this rectangle you might think of 7/10 as being well over a half, or even as not far from 1.

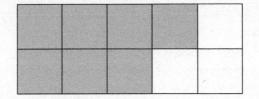

Rule Number 1 for working with fractions is to work only with simple fractions – ones you can visualize. If you come across a sum with a big-number fraction (e.g. 17/43) or a whole-and-fraction value (e.g. $5\,^3/_8$), convert the fraction to a decimal and use your calculator.

Same fraction, different numbers

There are often different ways of expressing the same fraction. For example, two quarters are the same as a half (2/4 = 1/2). Fractions are normally best expressed in their simplest form. You can simplify a fraction if you can divide both the top and bottom numbers by the same number, as in this example:

$$\frac{6 \div 2 = 3}{8 \div 2 = 4}$$

Practice 3

Simplify these fractions by dividing top and bottom by the same numbers – if possible.

1) 6/12
2) 8/10
3) 9/12
4) 12/16
5) 7/10

Adding fractions

You can't add different fractions together – not directly. This sum is not do-able:

1/2 + 1/3

Its fractions have different denominators (bottom numbers). However, if we express both fractions in terms of a common denominator, we can add them:

3/6 + 2/6 = 5/6

To find the common denominator, multiply the denominators. For example, to add 1/4 to 1/3, we would work in twelfths, because 4 × 3 = 12.

To convert two fractions before adding, we multiply both parts of each by the denominator of the other.

$$\frac{1}{4} + \frac{1}{3} = \frac{1 \times 3}{4 \times 3} + \frac{1 \times 4}{3 \times 4} = \frac{3}{12} + \frac{4}{12} = \frac{7}{12}$$

If the resulting fraction can be simplified, do so. 7/12 is as simple as it gets.

$$\frac{1}{2} + \frac{2}{3} = \frac{1 \times 3}{2 \times 3} + \frac{2 \times 2}{3 \times 2} = \frac{3}{6} + \frac{4}{6} = \frac{7}{6}$$

If the resulting fraction is top-heavy, convert it to an integer and a fraction. Here we can take 6/6 (= 1) away, leaving 1/6, to make 1¹⁄₆.

Subtraction

As with addition, you can only subtract fractions with the same denominator, so convert first if necessary. Multipy the top and bottom figures of each fraction by the denominator of the other. For example:

$$\frac{1}{2} - \frac{1}{3} = \frac{1 \times 3}{2 \times 3} - \frac{1 \times 2}{3 \times 2} = \frac{3}{6} - \frac{2}{6} = \frac{1}{6}$$

SUNDAY

MONDAY

TUESDAY

WEDNESDAY

THURSDAY

FRIDAY

SATURDAY

Practice 4

Try these. Simplify the result if you can.

1) 1/2 + 1/4

2) 1/4 + 2/3

3) 1/2 + 1/3 + 1/6

4) 3/4 + 1/2

5) 3/5 + 1/2

Multiplying by fractions

These are the most common sorts of sums involving fractions, and they are normally about sharing costs or other things. You know you've got a multiply-by-a-fraction job when it says 'of' in the middle, as follows:

1/4 of 18; 2/5 of 30; 3/4 of 240.

Written as sums, these are:

1/4 × 18; 2/5 × 30; 3/4 × 240.

And they are easy. Multiply by the top bit of the fraction and divide by the bottom bit – though it's often simpler to do the division first. So:

1/4 × 18 = 18 ÷ 4 = 4½

2/5 × 30 = 30 ÷ 5 = 6 × 2 = 12

3/4 × 240 = 240 ÷ 4 = 60 × 3 = 180

If the sum has fractions on both sides, e.g. a half of a quarter (1/2 × 1/4), multiply the top numbers then the bottom numbers. The result may need simplifying.

$$\frac{1}{2} \times \frac{1}{4} = \frac{1 \times 1}{2 \times 4} = \frac{1}{8} \qquad \frac{2}{5} \times \frac{3}{4} = \frac{2 \times 3}{5 \times 4} = \frac{6}{20} = \frac{3}{10}$$

It's easy – as long as you only do it with simple fractions. (Remember Rule Number 1.)

There's a limit!

In numeracy tests you should never have to do anything more complicated than halves, quarters, thirds, sixths and the like.

Dividing by fractions

If you are asked to divide by a fraction, you can usually deal with the problem in another way. For instance, the question 'If I buy seven pizzas and divide them into quarters, how many will that feed?' gives us this sum:

$7 \div 1/4$

However, if you pose the question as: 'If I buy seven pizzas and cut each into four, how many will that feed?' you get this sum:

7×4

which is a much nicer one (though the real answer is still zero – a quarter of a pizza won't feed anybody!).

If you are faced with a divide-by-fraction sum, deal with it this way. Turn the fraction upside down and multiply. Instead of:

$24 \div 3/5$

write:

$24 \times 5/3$

and work it out as follows:

$24 \times 5 = 120 \div 3 = 40$

As a rough check, remember that, if you are dividing by a value less than 1, you will get more, and the smaller the fraction, the bigger the end result will be. As you saw earlier, dividing by 1/4 is the same as multiplying by 4. If you divided the pizzas into 1/6, you'd get 6 slices per pizza.

Practice 5

Try these. Simplify the result if you can.

1) $2/3 \times 15$
2) $3/4 \times 32$
3) $27 \times 1/3$
4) $1/3 \times 1/2$
5) $1/4 \times 2/5$

Decimals

This next bit may be obvious, but there is a real point to it, so bear with me. Decimal fractions are an extension of the normal number system, where the value of a digit within a number depends upon its place. As you go left from the decimal point, each digit is worth 10 times more; as you go to the right, each digit is worth 10 times less. This means that, for example, 123.45 is worth:

1	2	3	.	4	5
×	×	×		×	×
100	10	1		1/10	1/100

What this means is that 0s can be crucial for establishing value. We are all pretty good at remembering them with integers, but we may not be so hot about remembering them on the other side of the decimal point. There's a lot of difference between 0.5 (1/2) and 0.005 (1/200).

How many decimal places?

Some multiplication and division sums can produce long strings of digits after the decimal point. In many cases, these are not necessary. If you are dealing in money, you would normally only bother with two digits after the point – the pence or cents. If you are a NASA scientist calculating a space flight, you would want it accurate to 20 or more decimal places, otherwise – at the end of your craft's 56-billion-mile journey – it might be 20 miles out and totally miss its target.

Adding and subtracting decimals

The key thing here is that you must line up the dots (the decimal points) in a column of numbers. Apart from that, addition and subtraction involving decimals is exactly the same as with integers.

For example:

123.45 + 56.789 + 32.1 + 0.007

Line them up!

```
 123.45
  56.789
  32.1
   0.007
= 212.346
```

You may also find it useful to add 0s to the ends so that they all have the same number of digits to the right of the decimal point – it's all too easy to miss a digit.

```
 123.450
  56.789
  32.100
   0.007
= 212.346
```

Quick check: 120 + 60 + 30 = 210. And that's near enough.

Check!

Checking is essential. One of the most common calculator errors is missing out the decimal point. We all mis-key occasionally, but you are much less likely to notice if you fail to connect with the dot key than with a number key – the dot may be just a single pixel on a pocket calculator.

The rule when checking sums involving decimals is: ignore the decimals unless they are more or less all you've got. For instance, if the sums involve money and most of the amounts are pounds and pence, ignore the pence, but if most of the sums are pence then round them to the nearest 10p.

Practice 6

Use your calculator to find the answers to these sums. In each case, check the answer with a nice round number estimate.

1) 555.34 + 12.895 + 107.107 + 8,765.76 + 0.002

2) 0.125 + 0.0125 + 0.000125   3) 1.99 + 4.99 + 12.49 + 0.25 + 0.33

4) 1.99 + 0.99 + 0.49 + 0.25 + 0.33   5) 1,000 − 234.56

Multiplying decimals

One of the things to remember about multiplying decimals is that they are fractions and, if you multiply a decimal (fraction) by another decimal (fraction), the answer will be smaller than either of them. No matter how big the numbers, if they have digits after the decimal point, then the answer will have a longer string of digits after the decimal point. There is a simple way to know how long that string will be, which gives you a way to find and/or check the answer.

The number of decimal places (d.p.) in the result will be the sum of the number of decimal places in each of the values being multiplied.

12.3 (1 d.p.) × 3.12 (2 d.p.) = 38.376 (3 d.p.)

Unfortunately, you cannot rely on this to check a calculator's result. If one of the numbers ends in 5 and the last digit of the other number is even, the result will end in 0 and the calculator won't display it. For example:

0.4 (1 d.p.) × 0.5 (1 d.p.) = 0.20 (2 d.p.)

but the calculator will show 0.2.

Checking the numbers

We need a better way of checking big and small numbers.

Big number check
If, in both numbers, the decimal fraction is just the tail end of a larger number, ignore the decimals.

Small number check
If one of the numbers is less than 1, here's what you need to do to get an estimate to check the result. Suppose you started with 1,234 × 0.0813.

1 Find the first significant digit (1 to 9), but make sure that you keep any leading zeros between it and the decimal point. Round it up if the next digit is 5 or more, and discard the remaining digits.

2 This means that 0.0813 becomes 0.08.

3 Count how many steps it would take to move the decimal point so that it is immediately to the right of the significant digit.

4 To get from 0.08 to 8.0 takes two steps.

5 Round the other number and multiply by the simplified decimal; 1,234 × 0.0813 thus becomes 1,200 × 8 = 9,600.

6 Move the decimal point back as many steps as you just moved it forward, so that 9,600 becomes 96.

7 The calculator gives us 100.3242 and 96 is the same order of value.

Here's an example where both numbers are less than 1.

- The sum: 0.1234 × 0.056
- Rounds to: 0.1 × 0.06
- Take the steps: 0.1 to 1.0 is one step; 0.06 to 6.0 is two steps = three steps in all.
- Significant digits: 1 × 6 = 6
- Step back: 6.0 step back three = 0.006
- Calculator result: 0.0069104

For a really rough check when both numbers are small, remember that a small bit of a small bit is going to be tiny.

Practice 7

Work out the estimate, and then use your calculator to find the exact answers. Does your estimated answer have the decimal point in the same place? Does it start with the same number or one very close?

1) 12.95 × 0.66

2) 153.33 × 0.0892

3) 55,532.19 × 0.0022

4) 0.5 × 0.25

5) 0.8 × 0.001

Dividing by decimals

Division is the mirror image of multiplication, so the same techniques apply – with minor differences. Use a calculator to get the answer. Check your answer with a rough estimate.

1 If both numbers are more than 1, ignore the decimal fractions and treat them as integers.
2 When you divide by a decimal fraction, the answer is going to be *bigger* than what you started with. At the simplest: 1 divided by 0.5 is the same as saying how many times will 0.5 go into 1, and the answer is 2.
3 Simplify the decimal to a nice round number, and count how many steps it would take to move the decimal point to its right.
4 Do the division as if they were both integers.
5 Add a zero for every step that you took at stage 3.

Here's an example:

● The sum: $1,234.56 \div 0.378$
● Calculator result: 3266.0317
● Rounds to: $1,200 \div 0.4$
● Take the steps: 0.4 to 4.0 takes one step
● Do the division: $1,200 \div 4$. $4 \times 3 = 12$, so $4 \times 300 = 1,200$
● Add the zeros: 300 + one step = 3,000 (which is the same size as the calculator result)

Practice 8

Use your calculator to find the answers to these sums. In each case, check the answer with a nice round number estimate.

1) $12.95 \div 0.66$

2) $35.13 \div 0.0775$

3) $524.78 \div 15.0022$

4) $0.5 \div 0.25$

5) $0.8 \div 0.001$

Summary

Today we looked at compound sums with series of calculations, and at fractions and how to do arithmetic with them. You learned the importance of the following points.

- Always follow the BODMAS order when calculating.
- Visualize fractions so that you have a clear idea of the size of the answer.
- If the top and bottom numbers in a fraction can be divided by the same value, do it. The simpler the fraction, the more likely you are to get the answer right.
- To add or subtract fractions, the denominators (bottom numbers) must be the same.
- If you multiply by a fraction or decimal that is less than 1, the answer will be smaller. If you divide by a value less than 1, the answer will be bigger.
- When you are dealing with decimals, keep a close eye on the position of the decimal point. For checking purposes, ignore the decimal fraction of any values greater than 1.

SUNDAY
MONDAY
TUESDAY
WEDNESDAY
THURSDAY
FRIDAY
SATURDAY

End-of-the-day test
[answers at the back]

Calculate or use your estimating skills to pick the right answer:

1) $12 + 6 \times 4 - 8 \times 2 + 7$

2) $7 + 6 \times 8 + 105$

3) $(7 + 6) \times 8 + 105$

4) $1/8 \times 2/3 =$
   a) 1/24   c) 2/21
   b) 1/12   d) 3/8

5) $2/5 + 1/4 =$
   a) 3/9   c) 13/20
   b) $1\frac{1}{2}$ d) 5/8

6) $3/4 \times 17 =$
   a) $12\frac{3}{4}$ c) $9\frac{1}{4}$
   b) 12/17 d) $21^3/_{17}$

7) $1/2 + 1/4 \times 3/4 =$
   a) 3/8   c) 11/16
   b) 13/8   d) 3/4

8) $3\frac{1}{4} + 12\frac{1}{2} + 6\frac{3}{4} + 2\frac{1}{5} =$
   a) $14^6/_{15}$ c) $21^6/_{10}$
   b) $15\frac{1}{4}$ d) $24^7/_{10}$

9) $8.35 \times 0.06 =$
   a) 0.5118 c) 5.301
   b) 5.3   d) 0.501

10) $1{,}234.56 + 67.89 + 0.043 + 3.5 =$
    a) 1,303.23   c) 8,027.103
    b) 1,305.993 d) 1,350.93

Find the answer. Calculators are allowed here.

11) $75 \times 6 + 35 = ?$
    a) 116   c) 485
    b) 3,075 d) 503

12) $(3 \times 7) - 12 - (6 \times 2) + 3 = ?$
    a) – 3   c) 3
    b) 0   d) 6

13) $13\,^3/_4 \times 27\,^2/_5 =$

14) $2/5 \times 3/7 \times 5/8 \times 14/15 =$

15) $7/25 \times 264 =$

16) $538.74 + 76.96 =$

17) $538.74 \div 76.96 =$

18) $5.6 + 7.2 \times 12.5 =$

19) $(4.99 + 1.49 + 2.25) \times 3.5 =$

20) $67.2 \div (14.7 - 3.5) =$

True or false?

21) $8.5 + 13.75 - 2.25 = 20$

22) $3/5 + 3/10 + 3/20 = 19/20$

23) $7/10 - 2/5 + 1/15 = 2/5$

24) $0.002 \times 5{,}600 = 11.2$

25) 15% of $300 = 45$

26) $1/2 + 1/4 \times 3/4 = 3/8$

27) $853.57 - 214.22 = 639.35$

28) $\$4{,}560.63 \div \$27.05 = \$618.80$

29) $13\,2/5 \times 55 = 737.0$

30) £$5{,}742.30 \times 16 =$ £$9{,}876.80$

Answers to practice questions

Practice 1: 1) 8½ ; **2)** 6½ ; **3)** 42; **4)** 68; **5)** 13.

Practice 2:

1) 865. Estimate: 200 + (25 × 10) + 400 = 850.

2) 35. Estimate: 90 – 20 = 70. 2,000 ÷ 70 = 200 ÷ 7 = a bit under 30, but 2,000 is a lot under 2,485, so the answer will be something over 30.

3) 13,688. Estimate: Brackets first, 500 + 100 + 200 = 800 × 20 = 16,000 + 300. But it was 16, not 20, so quite a bit less than 16,300.

4) 243. Estimate: 4 × 70 = 280, call it 300; 20 × 3 = 60. 600 – 300 – 60 = 240.

5) 12. Estimate: 70 – 20 = 50 × 4 = 200 × 3 = 600. 600 – 600 = 0.

Practice 3:

1) 6/12 = 1/2

2) 8/10 = 4/5

3) 9/12 = 3/4

4) 12/16 = 3/4

5) 7/10 = 7/10

Practice 4:

1) 1/2 + 1/4 = 2/4 + 1/4 = 3/4

2) 1/4 + 2/3 = 3/12 + 8/12 = 11/12

3) 1/2 + 1/3 + 1/6 = 3/6 + 2/6 + 1/6 = 6/6 = 1

4) 3/4 + 1/2 = 3/4 + 2/4 = 5/4 = 1 1/4

5) 3/5 + 1/2 = 6/10 + 5/10 = 11/10 = 1 1/10

Practice 5:

1) 2/3 × 15 = 2 × 15 ÷ 3 = 30 ÷ 3 = 10

2) 3/4 × 32 = 3 × 32 ÷ 4 = 96 ÷ 4 = 24

3) 27 × 1/3 = 27 ÷ 3 = 9

4) 1/3 × 1/2 = 1 × 1 ÷ 3 × 2 = 1 ÷ 6 = 1/6

5) 1/4 × 2/5 = 1 × 2 ÷ 4 × 5 = 2 ÷ 20 = 2/20 = 1/10

Practice 6:

1) 555.34 + 12.895 + 107.107 + 8,765.76 + 0.002 = 9,441.104 (500 + 0 + 100 + 9,000 = 9,600 + 0 = 9,700 – check!)

2) 0.125 + 0.0125 + 0.000125 = 0.137625 (0.1 + 0.01 + 0.001 = a bit more than 0.11 – check!)

3) 1.99 + 4.99 + 12.49 + 0.25 + 0.33 = 20.05 (2 + 5 + 12 = 19 and a bit more – check!)

4) 1.99 + 0.99 + 0.49 + 0.25 + 0.33 = 4.05 (2 + 1 + 0.5 + 0.3 + 0.3 = 4.1 – check!)

5) 1,000 – 234.56 = 765.44 (1,000 – (a bit more than) 200 = (a bit less than) 800 – check!)

Practice 7:

1) $12.95 \times 0.66 = 8.547$ ($13 \times .7 = 13 \times 7$ (one step) $= 91$, step back to 9.1 – check!)

2) $153.33 \times 0.0892 = 13.677036$ ($150 \times 0.09 = 150 \times 9$ (two steps) $= 1,350$, step back to 13.5 – check!)

3) $55,532.19 \times 0.0022 = 122.170818$ ($60,000 \times 0.002 = 60,000 \times 2$ (three steps) $= 120,000$, step back to 120 – check!)

4) $0.5 \times 0.25 = 0.125$ ($0.5 \times 0.3 = 5 \times 3$ (two steps) $= 15$, step back to 0.15 – check!)

5) $0.8 \times 0.001 = 0.0008$ (8×1 (four steps) $= 8$, step back to 0.0008 – check!)

Practice 8:

1) $12.95 \div 0.66 = 19.621212$ (0.66 becomes 0.7 becomes 7 in one step; $12 \div 7 =$ almost 2, move a step $=$ almost 20)

2) $35.13 \div 0.0775 = 453.29032$ (0.0775 becomes 0.08 becomes 8 in two steps; $35 \div 8 = 8 \times 4 = 32$, so $35 \div 8 =$ something over 4, move two steps $=$ something over 400)

3) $524.78 \div 15.0022 = 34.980202$ (round both to $500 \div 20 = 25$. Because 15 is quite a lot less than 20, the answer will be quite a bit more than 25)

4) $0.5 \div 0.25 = 2$ (move the decimal point to the left in both of them: $5 \div 2.5$ or $50 \div 25 = 2$)

5) $0.8 \div 0.001 = 800$ (move the decimal point to the left: $8 \div 0.01$ or $80 \div 0.1 = 800 \div 1 = 800$)

SUNDAY

MONDAY

TUESDAY

WEDNESDAY

THURSDAY

FRIDAY

SATURDAY

TUESDAY

Percentages, ratios and averages

Today we look at three more areas of arithmetic: percentages, ratios and averages. With these we will have a full toolkit that we can use when tackling a wide range of numerical problems, whether they are given as straightforward sums – such as we've had so far – or in the form of stories, of 'what comes next?' sequences, or of table- or graph-based data interpretation problems.

- A percentage is a type of fraction: 70% means 70/100. And percentages can sometimes be converted to simple fractions to give you easier sums. Three obvious ones are 10% = 1/10; 25% = 1/4; 33.3% = 1/3.
- Ratios are about the relationship between values. They are central to conversion-based problems, which are popular in numeracy tests.
- Averages can be defined in three ways: mode, median and mean, but the only one normally used in testing is the mean – the numerical average.

What's a percentage?

The clue is in the name: 'per cent' is Latin for 'in a hundred' or 'for every hundred'. Thus '50 per cent' means '50 for every 100'. If a discount is 25%, it means that they will reduce the price by 25p in every £1 (100p). If your Council Tax has gone up 10%, then for every £100 you paid last year, this year you will be paying an extra £10.

Percentages can also be used to express multiples. For example, 200% means 200 for every 100, or twice as much; 1,000% is 1,000 for every 100, or 10 times as much.

What's 15% of...?

The most common type of percentage calculation that most of us have to do is to find a given percentage of an amount: a 10% service charge on a bill, a 25% discount in a sale. These are straightforward to do, and there are some shortcuts that will make some calculations even simpler.

The rule is: **multiply the base amount by the percentage number and divide by 100.**

For instance:

- 5% of £30 = 30 × 5 ÷ 100 = 150 ÷ 100 = £1.50
- 25% of 80 = 80 × 25 ÷ 100 = 2,000 ÷ 100 = 20

Shortcuts

If you want 25% of something, you can multiply it by 1/4 or – more simply – divide it by 4. Here are some shortcuts worth noting:

Percentage	Fraction	Sum to do
10	1/10	÷ 10
20	2/10 = 1/5	÷ 5
25	1/4	4
33 1/3	1/3	÷ 3
50	1/2	÷ 2

Don't forget that, if the percentage is more than 100%, the result will be bigger than the base figure. For example, if you borrowed £200 for a year from a high street bank at 8%, it would cost you:

$8 \times 200 \div 100 = 8 \times 2 = £16$

If you borrow the same amount from a payday loan company at 4,000% (which is what their rates can work out at on an annual basis), it would cost:

$4,000 \times 200 \div 100 = 4,000 \times 2 = £8,000$

Practice 1

Use your calculator only if you have to, but always check the answer with an estimate.

1) 25% of 500   3) 17.5% of $400   5) 150% of 88

2) 300% of £54   4) 50% of (40% of £12.00)

What's that as a percentage?

Another common type of percentage problem is to express one value as a percentage of a second, e.g. 'What is 15 as a percentage of 60?' To find the answer, divide the first by the second and multiply by 100:

$15/60 \times 100 = 0.4 \times 100 = 40\%$

When checking your answer, the first-level check is this: if the first number is smaller than the second, the percentage will be less than 100. This may seem ridiculously obvious, but it picks up mis-keys on calculators. If it passes that test, use nice round numbers to estimate the answer.

Practice 2

Use your calculator if you have to, and check the answer.

1) What is 225 as a percentage of 300?

2) What is 45 as a percentage of 180?

3) What is 120 as a percentage of 60?

4) What is £14.60 as a percentage of £78.00?

5) What is $0.50 as a percentage of $23.50?

Ratios

A ratio is a way of expressing the relationship between two or more numbers. Ratios can be written like this: 10 : 3 or 5 : 2 : 1 with the numbers separated:by:colons. Where there are only two numbers, the ratio can be written as a single value, e.g. 0.85, which is equivalent to 1 : 0.85.

It's important to remember that ratios have no units. The perfect martini ratio is 3 : 1. You could mix 3 capfuls of gin to 1 capful of vermouth, or 3 glasses of gin to 1 glass of vermouth, or use any other unit as long as the ratio stays the same. So if you started with a bottle of gin – 75 cl – you would need 1/3 of that – 25 cl – of vermouth. (You'd also need 2 litres of ice cubes, a 3-litre shaker and a strong bartender.)

Calculations using ratios

Working with ratios is similar to working with fractions. In fact, you can express any (whole number) ratio as a set of fractions – just add up all the numbers to get the denominator. For example, with 3 : 1 gin to vermouth, 3 + 1 = 4 and the recipe could be expressed as 3/4 gin and 1/4 vermouth.

Concrete is a mixture of cement, sand and gravel in the ratio 1 : 2 : 4. Therefore, if some builders needed 12 cubic metres of concrete for foundations, how much would they need of each?

1 First add the ratio numbers to find how many parts there are: $1 + 2 + 4 = 7$.
2 Then divide 12 by 7 to get the size of one part: $12 \div 7 = 1.714$ cubic metres (m^3) (that's the cement).
3 Then multiply to find the other values:
   a) $2 \times 1.714 = 3.428$ cubic metres (sand)
   b) $4 \times 1.714 = 6.857$ cubic metres (gravel).

Sticking with the concrete, let's explore the ratio from the other end. If you've got 18 bags of gravel, how much sand and cement will you need? To work this out, first divide 18 by 4 to get 4.5. Now multiply the other parts by this:

$1 \times 4.5 = 4.5$ bags cement

$2 \times 4.5 = 9$ bags sand

Practice 3

1) Share 60 in the ratio $2 : 3$
2) Split 80 in the ratio $1 : 3 : 4$
3) Find the missing value: $4 : 6 = y : 30$
4) Find the missing values: $2 : 4 : 8 = y : \$14 : z$
5) Find the missing values: $3 : 1 : 5 = y : z : £62.50$

Conversions

Most conversions are done in basically the same way, because you are doing basically the same thing – changing the units that you use to describe a quantity. The quantity may be length, weight, volume, value or whatever, and the units of measurement may be different, but the process is the same. To do a conversion you need to know two things:

1 **Which unit of measurement is bigger**, and roughly by how much. For instance, 1 litre is about 2 pints. Use this to do a rough check on the answer. If you started with litres, you should have about twice as many pints. If you started with pints, you should have about half as many litres. The relative size also tells you what sort of sum you have to do. To go from litres to

pints, you need to multiply to get a bigger number. To go from pints to litres, and get a smaller number, you need to divide.

2 **The magic number, or conversion rate**. For example, to convert inches to millimetres (mm), or vice versa, the magic number is 25.4 because 25.4 mm = 1 inch.

The fly in the ointment

I've been talking about *units of measurement*, but there are also *systems of measurement*. Sometimes it's not enough to know the magic number to convert miles to kilometres. If you start with mixed units – miles and yards – there's no single magic number. In these situations:

1 Convert the mixed units to one unit, e.g. miles and yards to yards – and to do this you need to know how many yards there are in a mile.

2 Convert the quantity in the single unit to the other system, e.g. yards to metres.

Practice 4

1) If £1 = €1.25, how much is £300 in euros?

2) If £1 = €1.25, how much is €1,000 in pounds?

3) 1 mile = 1.6 km. 621 km = ? miles

4) 1 m² = 10.76 ft². 35 m² = ?? ft²

5) Convert 8 st 5 lb to kilograms. 14 lb = 1 st; 1 lb = 0.454 kg.

Averages

If they don't say otherwise, when people talk about 'the average', they mean the 'arithmetical mean'. This is the number you get by adding up all the values in a set and dividing them by the number of values.

For example, the average of

2, 3, 4, 4, 6, 8, 8, 9, 10, 12
$= (2 + 3 + 4 + 4 + 6 + 8 + 8 + 9 + 10 + 12) \div 10$
$= 66 \div 10 = 6.6$.

As a very rough check, count how many values are above the average and how many are below it (and allow some leeway if a value is way above or below). If they are more or less the same, your calculation is probably right. Does the check work here?

Here's another example, and one that brings out a key point. Last week I jogged 6 miles on Monday, 4 on Tuesday, 5 on Wednesday, 6 on Thursday, 0 on Friday, 4 on Saturday and 10 on Sunday. What was my daily average for the week? The calculation is:

$$(6 + 4 + 5 + 6 + 0 + 4 + 10) \div 7 = 35 \div 7 = 5 \text{ miles}$$

Notice that Friday is included in the count of days, even though I didn't run that day, because the question asked for the daily average. If the question had been 'What is the average length of each run?', the calculation would have been:

$$(6 + 4 + 5 + 6 + 4 + 10) \div 6 = 35 \div 6 = 5.83 \text{ miles}$$

Which numbers count?

When data is collected, there may be zero values, and these may or may not be included in the count. It depends upon what is being measured and why. As so often with numbers, if there is confusion, it comes from the words around the numbers.

Practice 5

1) Find the average of: £12.35, £48.60, £9.72, £28.25, £16.40.
2) Find the average of: 7, 3, 9, 2, 5, 8, 6, 7, 4, 8.
3) Find the average of: 345.6, 736.5, 93.7, 123.68, 546.87, 88.54.
4) An apple tree produces 20 fruits of 100 g, 30 of 110 g, 15 of 120 g and 5 of 125 g. What is the average weight (to the nearest gram) of its apples?
5) A firm pays 12 salaries of £15,000, 25 of £25,000, 4 of £35,000 and 3 of £48,000. What is the average salary?

Summary

Today we looked at some more arithmetical operations.

- Most percentage sums are multiplication. Simply multiply by the percentage number and divide by 100 to get the answer. If you have over 100%, the result will be bigger than the start value.
- When working with ratios, if you know how many parts there are and the total value, you can work out the value of each part. Where values are not given, it can help to drop in some nice round numbers so you have something real to work on.
- The first stage in any conversion is to note which unit is larger, so that you know whether you should have more or fewer after converting the numbers. If you start with mixed units (e.g. feet and inches), convert them to the smallest unit.
- When a question asks for the average, it usually wants the arithmetic mean – the sum of all the values divided by how many there are.

SUNDAY
MONDAY
TUESDAY
WEDNESDAY
THURSDAY
FRIDAY
SATURDAY

End-of-the-day test

[answers at the back]

Select the correct answer:

1) 15% of 400 =
   a) 15   c) 60
   b) 45   d) 90

2) 125% of $24.80 =
   a) $31.00   c) $18.60
   b) $6.20   d) $3,100.00

3) 17.5% of £35.50 =
   a) £2.67   c) £53.00
   b) £6.21   d) £8.75

4) 2 : 7 = ?? : 35
   a) 8   c) 14
   b) 10   d) 16

5) 3 : ?? = 27 : 180
   a) 15   c) 20
   b) 18   d) 21

6) 1 mile = 1.6 km;
   225 miles = ?? km
   a) 360   c) 288
   b) 320   d) 248

7) 1 kg = 2.2 lb; 62.5 kg = ?? lb
   a) 162.5   c) 135
   b) 126.5   d) 137.5

8) The average of 12, 15, 23, 16,
   15, 21, 9, 17 =
   a) 10   c) 16
   b) 14   d) 17.5

9) The average of 5.6, 1.9, 4.3, 6.2,
   5.9, 1.8, 4.9, 7.2, 8.4, 4.8 =
   a) 5.1   c) 6.0
   b) 5.4   d) 6.5

10) The average of 443, 672, 0, 59,
    418, 349, 72, 785 =
    a) 325.14   c) 346.25
    b) 325.75   d) 349.75

You may use a calculator here to find the correct answer:

11) 250% of £349.50 =

12) 15% of (24 × $3.70) =

13) (£6.26 + £7.50 + £3.50 + £0.99) × 12.5% =

14) Share 98 in the ratio 1 : 2 : 4.

15) Share £2,450,650 in the ratio 3: 2: 2: 4.

16) Convert 14 feet 7 inches to metres. 1 inch = 2.54 cm; 12 inches = 1 foot.

17) Convert $500 to euros, at the rates $1 = £0.62 and €1 = £0.79.

18) How many 6-inch square tiles are needed to cover a bathroom wall 2.4 m long to a height of 1.3 m?

19) In a firm, the boss pays himself £100,000, his two managers get £40,000 and the 25 workers earn £18,000. What is the average salary?

20) What is the average shoe size in this group of children?

Number	4	8	12	7	2
Size	3	4	5	6	7

Answers to practice questions

Practice 1:

1) 25% of 500 = 125 (25% is 1/4. 500 ÷ 4 = 125)

2) 300% of £54 = £162 (300% = × 3)

3) 17.5% of $400 = $70 (round to 20% gives 20 × 400 ÷ 100 = 80)

4) 50% of (40% of £12.00) £2.40 (work out the brackets first: 40% of £12 = 40/100 × 12 = 4/10 × 12 = 48/10 = 4.8; then work out 50% × £4.80 = 1/2 × £4.80 = £2.40)

5) 150% of 88 = 132 (rounds to 1½ × 90 = 90 + 45 = 135).

Practice 2:

1) 75%; 2) 25%; 3) 200%; 4) 18.72%; 5) 2.13%.

Practice 3:

1) The shares are 24 and 36 (2 + 3 = 5, so 5 parts in total. 60 ÷ 5 = 12. 2 × 12 = 24, 3 × 12 = 36)

2) The shares are 10, 30 and 40. Use the same process as in Q1.

3) $y = 20$. 30 ÷ 6 = 5, so multiply 4 by 5 to get the same proportion.

4) $y = \$7$, $z = \$28$. 14 ÷ 4 = 3.5; 2 × 3.5 = 7; 8 × 3.5 = 28.

5) $y = £37.50$, $z = £12.50$. 62.50 ÷ 5 = 12.5; 3 × 12.5 = 37.5.

Practice 4:

1) €375 (rough check: euros are smaller than pounds, so there will be more of them)

2) £800

3) 388.13 miles (miles are bigger than kilometres, so there will be fewer of them for the same distance)

4) 376.6 ft^2

5) 8 st 5 lb = 8 × 14 + 5 = 117 lb × 0.454 = 53.12 kg

Practice 5:

1) £23.06 (the rough check shows three under this and two over it, but one is quite a lot over, so it's probably right. Doing the sum with round numbers gives you 10 + 50 + 10 + 30 +20 = 120 ÷ 5 = 24)

2) 5.9

3) 322.48

4) First multiply the numbers by each weight and add to find the total weight bill, then find the total number of fruit, then divide ((20 × 100) + (30 × 110) + (15 × 120) + (5 × 125)) ÷ (20 + 30 + 10 + 15) = 7,725 ÷ 70 = 110 g)

5) £24,750 (use the same techniques as for Q4: (12 × 15,000) + (25 × 25,000) + (4 × 35,000) + (3 × 48,000) ÷ (12 + 25 + 4 + 3) = 1,089,000 ÷ 44 = 24,750.

WEDNESDAY

Sums from stories

The main focus today is on how to get sums from stories. When a problem is presented in words, not just numbers, what sort of calculation is required? What are the relevant numbers and which parts of the story can be ignored?

Prospective employers don't simply want to know that you can do sums quickly and accurately. If you get the job, you are going to have to use those skills in the context of work, and employers also need to know that you know what sort of sums to do in a given situation.

Today we will look at:

- how to express problems as numbers
- how to know which numbers are relevant
- how to simplify a problem
- how to sketch it
- some useful formulae.

Making sense of the story

When a problem is presented in words, not just numbers, you need to make sure that you get from it the sum, the whole sum and nothing but the sum.

Simple sums

The sum may be obvious and very simple. For example:

Q1. What is the area of a rectangular room 4.2 m by 2.7 m? This takes you directly to 4.2 × 2.7.

This sum may be obvious, but you may be presented with a more complicated question, which can indicate a more complex operation or series of operations. For example:

Q2. A room is 4.2 m by 2.7 m and 2.1 m high. How many pots of paint will be needed for the walls if a pot will cover 6 m^2? The door and window occupy a total of 3.5 m^2.

In this case you'll need to take the following steps.

1 Calculate the perimeter of the room: (4.2 + 2.7) × 2 = 13.8 m.
2 Multiply the perimeter by the height to get the overall wall area: 13.8 × 2.1 = 28.98 m^2.
3 Subtract the door and window area: 28.98 − 3.5 = 25.48 m^2.
4 Divide by 6 to find the number of pots – rounding up to a whole pot: 25.48/6 = 4.25 pots, which rounds up to 5 pots.

Which numbers are relevant?

The problem may contain distracters – numbers that are not needed, and that are there to test your ability to see what is relevant. Don't be distracted. Ask yourself these two questions:

● What am I being asked to find out?
● What do I need to know to find it?

For example:

Q. A car going from Brighton to London covers 34 miles in 30 minutes. If it is 55 miles from Brighton to London, how fast is the car travelling?

You are being asked to find the speed – the mph. To work that out you need to know the distance covered and the time taken. In this question, the distance covered is 34 miles and the time is 30 minutes. The total distance from Brighton to London is irrelevant.

Practice 1

Express these problems as sums and find the answers. Use a calculator if you need to, and always check each answer with a round number estimate.

1) The bill in a restaurant lists £12.50 (for starters), £25.88 (for mains) and £15.40 (for drinks). With a 10% tip, what is the total cost?

   a) £59.16   b) £61.28   c) £55.14   d) £57.75

2) If 8 machines make 15,000 widgets a week, how many can 4 machines make in 4 weeks?

   a) 60,000   b) 25,000   c) 32,000   d) 30,000

3) A room is 12 feet 5 inches by 9 feet 7 inches. What is the floor area in square metres (m²)? Assume 1 inch = 2.5 cm.

4) What's the cost of 43.57 litres of petrol at 115.35p a litre? Answer in £s, please.

5) You've spent 11 hours 45 minutes on a job for a client and your time is charged at £75.80 an hour. What will your time cost?

Practice 2

Express these problems as sums and find the answers. Use a calculator if you need to, and always check each answer with a round number estimate.

1) A garden 11 m long and 5 m wide needs to be fenced all the way round. How many 3-m by 1.8-m panels will be needed?
2) George earns £45,000 p.a. and has been with the same firm half his life. Sarah is 52 and only joined the firm six years ago. If Sarah is four years older than George, how long has he been with the firm?
3) The Smiths drive to the seaside in their 995 cc Citroen, taking 2 hours 20 minutes, with a 10-minute stop half way. Their neighbours, the Joneses, leave 15 minutes later, in their 3-litre Rover, drive non-stop and arrive at the seaside 5 minutes before them. What is the ratio of the speeds of the Smiths' and Joneses' cars?
4) A school has 1/4 fewer computers than pupils, and two printers per classroom. If there are currently 80 pupils and 10% more arrive, how many more computers will be needed?
5) A box of 25 rolls of tape costs £11.80, a box of 50 costs £22.85, a box of 75 costs £30.15 and a box of 100 costs £42.10. Which rolls are the cheapest?

Simplify the problem

Some problems look more complicated than they are. If you can't see a way to answer what seems to be the obvious question, perhaps you are asking the wrong question.

For example:

Q. A boat travels upstream for 4 hours 24 minutes, and then returns downstream to its start point in 2 hours. What is the ratio between the speed of the boat and speed of the current?

You can't work out the speed here, because the distance is not known. All you have is time. However, the ratio of the speeds will be equivalent to the time elapsed, and we can work that out.

In still water, the boat would have covered the distance in the average time of the upstream/downstream trips = (4 h 24 min + 2 h)/2 = 3 h 12 min. It is the current that causes the difference between this and the times of the trip (either way).

4 h 24 min – 3 h 12 min = 1 h 12 min.

The ratio is 3 h 12 min: 1 h 12 min = 192 min: 72 min = 8:3.

Practice 3

1) John takes 1 hour 15 minutes to drive to his mother's for Sunday lunch. Hurrying home to catch the big match on TV, he drives 50% faster than on the way up. How long does his return journey take?

2) You've been asked to organize the firm's picnic for 33 people. You've found a caterer who will supply food (and crockery/cutlery) at £8.50 a head. Glass hire is £5.25 a dozen (people can manage with one each). You'll need four crates (each of a dozen) of wine at £5.99 a bottle, plus 8 litres of fruit juice at £2.45 per litre. What is the total cost?

3) The hospitality fund will cover the first £250 of the cost of the picnic. How much will each of the 33 participants have to pay?

Sketch it

You may be presented with problems in the form of a figure, to be solved with some elementary geometry. For example:

Q. The end wall of a garage is to be coated with a weatherproofing paint. If two litres of paint cover one square metre, how many litres are needed? The wall's dimensions are shown in this sketch:

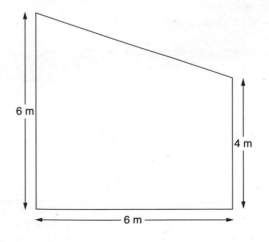

With these kinds of problems, you should split the shape into simple components – rectangles, triangles and circles – and then work out the areas and add or subtract as necessary. In this case, split it into a rectangle and a triangle, and use the following formulae.

- **Area of a rectangle = base × height (or length × width)**
- **Area of a triangle = (base × height) / 2.**

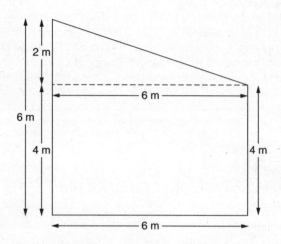

$6 \times 4 + (6 \times 2) \div 2 = 24 + 6 = 30 \text{ m}^2$

Therefore 60 litres of paint are needed.

Other useful formulae	
Volume of a block	length × width × height
Perimeter of a rectangle	(base + height) × 2
Area of a circle	πr^2 or pi × radius × radius
Perimeter of a circle	$2\pi r$

 In a calculated sum, take pi as 3.141; for estimating, use 3.

Remember Pythagoras!

In a right-angled triangle, the square on the hypotenuse (the angled side) = the sum of the squares on the two other sides.

There are some convenient whole-number triangles that they like to use in tests. The most common is 3:4:5 ($3^2 + 4^2 = 5^2$ or $9 + 16 = 25$) and multiples of this such as 6:8:10 and 9:12:15. Less often, you will meet 5:12:13 and 8:15:17.

Practice 4

1) What is the area of this field?

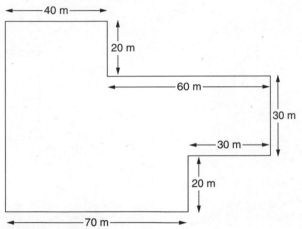

2) Sue goes orienteering and runs 5 km south, then 2 km west, 2 km north and 6 km east. How far is she from the start, in a straight line?

3) In the middle of an 8-m square lawn is a circular pool 4 m in diameter. What is the area of the grass?
   a) 13.74 m²   b) 32.2 m²   c) 51.44 m²   d) 57.62 m²

Summary

Today you learned that when a problem is presented in words, not just numbers, you need to make sure that you get from it the sum, the whole sum and nothing but the sum.

Remember the following key points:

- The problem may contain distracters – numbers that are not needed and are there to test your ability to see what is relevant.
- Some problems appear to be more complicated than they are. If you don't have the numbers to solve what you think is the problem, try it from the other end. Start from the numbers and ask yourself what problem they could solve.
- Problems about area, distance, position and similar can usually be solved more easily with the help of a diagram.

SUNDAY
MONDAY
TUESDAY
WEDNESDAY
THURSDAY
FRIDAY
SATURDAY

End-of-the-day test

[answers at the back]

1) You've been asked for a list of employees broken down by age and sex. There are 49 people in total. Of these, 32 are female and 17 are over 50. What percentage are male? What percentage are over 50?

2) Sam's car weighs 470 kg empty but 850 kg when his whole family and their luggage are on board. The car can carry up to 80% of its weight. Is it overloaded?

3) Uncle Jasper specified in his will that his estate was to be split between his three ex-wives in the ratio 5:11:2 – these being the number of years each put up with him. He left £2.7 million. How much did each ex get?

4) My moped has a two-stroke engine that runs on a mixture of petrol and oil in the ratio 40:3. When I fill the tank with 5 litres of petrol, how much oil should I add?

5) Jake, Bart, Sam and Nat share the loot from the bank job in the ratio 5:3:2:6 respectively.If Jake gets $9,000 more than Sam, how much did they steal in total?
   a) $30,000   c) $48,000
   b) $42,000   d) $72,000

6) The proportion of boys to girls at a school is currently 6:5. If the number of boys increases by 10% and of girls by 20%, what will the boy:girl ratio be then?
   a) 5: 6   c) 10:11
   b) 11:10   d) 11:12

7) A plane takes 6 hours to fly 1,800 miles into a 100 mph headwind. How long will the return journey take if the plane's speed and wind speed are the same?
   a) 3 h 36 m   c) 4 h 24 m
   b) 4 h 10 m   d) 4 h 30 m

8) In January 2000 I converted $5,000 into euros. The exchange rate was then $1.00 = £0.70 and €1 = £0.63. In July 2012 I converted it back into dollars, when the exchange rates were $1.00 = £0.61 and €1.00 = £0.85. How much did I gain or lose?
   a) – $2,357.34   c) + $2,741.35
   b) – $1,770.79   d) + $3,216.57

Answers to practice questions

Practice 1:

1) a (£53.78 + £5.38 tip)

2) d (4 machines can make 7,500 widgets a week, times 4 for the 4-week total)

3) Convert the lengths to metres first. 12 ft 5 in = 149 in = 149 × 2.5 = 372.5 cm = 3.725 m; 9 ft 7 in = 115 in = 115 × 2.5 = 287.5 cm = 2.875 m. Area = 3.725 × 2.875 = 10.71 m².

4) 43.57 × 115.35 = 5,025.7995p = £50.26. The estimate would be a bit over 40 × a bit over £1, giving an answer of a fair chunk over £40, which is what we have.

5) 11.75 × 75.80 = £890.65. The estimate is a bit over 10 × a chunk under £80, giving something around £800, which we have.

Practice 2:

1) 12 panels. The fence needs four for each side and two for each end. Panels can be cut to length, but you wouldn't use the offcuts. The 1.8 m height is a distracter.

2) 24 years. To solve this, you need to work out George's age. He's 4 years younger than Sarah, who is 52. His salary and how long she has been with the firm are irrelevant.

3) The Smiths' travelling time is 2 hr 10 mins (or 130 mins). The Joneses' travelling time is 2 hrs (120 mins). The ratio is 130:120 or 13:12. The engine sizes are distracters.

4) Six computers. The printers are the first distracter and should be ignored. The second distracter is the phase '1/4 fewer'. A quick read of the question could leave you thinking one computer for every four pupils, when it's actually three computers. 10% of 80 = 8, and 8 pupils need 6 computers.

5) The 75 box. The distracter here is that you may be led into doing more calculations than necessary. You do not need to work out the cost per roll in every box. Twice £11.80 (25 box) is £23.60, which is more than £22.85 (50 box), and twice £22.85 is £45.70 more than £42.10 (100 box). You only need to work out the roll costs for the 75 and 100 boxes. £30.15 ÷ 75 = £0.402 and £42.10 ÷ 100 = £0.421.

Practice 3:

1) It helps to have some numbers. His outbound journey takes 75 minutes, so assume it is 75 miles. That makes his speed 60 mph. At 50% faster, he would be travelling at 90 mph and would cover the distance in 75/90 h = 50 min. (You could have assumed any distance – this just gives the simplest sums.)

2) The sum here is: (33 × £8.50) + (£5.25 × 3) + (4 × 12 × £5.99) + (8 × £2.45). The brackets mark the sections that you should add into memory and check with estimates. Here's how they should work out:

Section	Subtotal	Estimate
33 × 8.50	280.50	30 × 10 = 300
5.25 × 3	15.75	5 × 3 = 15
4 × 12 × 5.99	287.52	4 × 12 = 48. Call it 50 × 6 = 300
8 × 2.45	19.60	10 × 2 = 20
	Total = 603.37	Estimate 635

3) 603.37 – 250 = 353.37. Divided by 33 = £10.71. The estimate here is 300 ÷ 30 = 10.

Practice 4:

1) 5,200 m². Here's one way to tackle this. Treat it as one large rectangle with two smaller ones cut out of it. Add up the lengths to work out the size of the large rectangle.

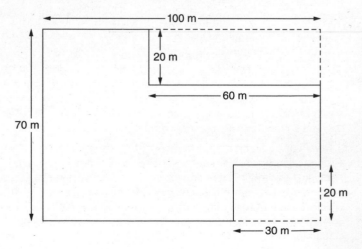

The area is then:

$(100 \times 70) - (60 \times 20) - (30 \times 20) = 7{,}000 - 1{,}200 - 600 = 5{,}200.$

2) 5 km. Draw her movements, and you will find that she is 3 km south and 4 km east of home. The direct line back is the hypotenuse of a 3:4:5 triangle.

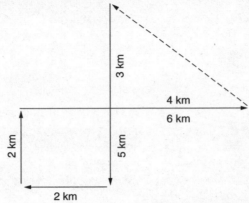

3) c. The grass area is the lawn minus the pool area. Try to draw a sketch more or less to scale.

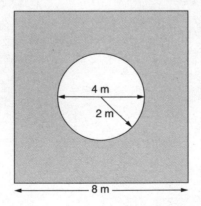

The pool radius = half the diameter = 2 m. A rough estimate, based on the sketch, is that the grass is over half the area of the lawn, which rules out (a) and (b). With pi as 3, we get $8 \times 8 - 3 \times 2 \times 2 = 64 - 12 = 52$. (c) is just under this, so will be the answer. On a calculator, $8 \times 8 - 3.141 \times 2^2 = 64 - 12.56 = 51.44$.

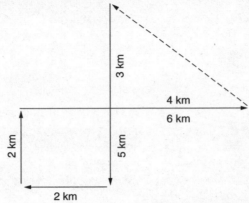

THURSDAY

Sequences and number patterns

In numerical tests, problems based on series, sequences or number patterns are designed to test your feel for numbers and your ability to spot relationships between them. Having a good mathematical brain helps here, but knowing the ways in which sequences and patterns can be made helps even more – and you can learn that.

Today we will look at:

- simple sequences, where each number is related directly to the next
- second-level sequences, where you have to find the sequence in the steps between the terms
- interwoven sequences, with two or more sets of numbers alternating
- alphabetical sequences, where letters are used in place of numbers
- patterns of numbers, where you are asked to find the relationship between sets of numbers.

In most problems numbers should be treated as values, and you will be performing some kind of arithmetical operation on them. Sometimes you need to look at the individual digits that make up a number and manipulate them in other ways.

Simple sequences

In the simplest sequences, the same operation is carried out at each step. It may be addition, subtraction, multiplication or division, but using the same value. These sequences can be recognized within the first two or three steps.

1, 4, 7, 10, 13 ...

The numbers here go up in steps of 3. You will have worked this out by the time you reach 7, and confirmed it by 10.

4, 8, 16, 32 ...

This time the operation is × 2 and, again, it is visible by the time you get to the third value.

Practice 1

1) 52, 48, 44, ...

2) 24, 41, 58, ...

3) 81, 27, 9, ...

4) 2.5, 4.0, 5.5, ...

5) 143, 197, 251, ...

Second-level sequences

If you apply an operation not to the numbers in the series but to the differences between the numbers, you get a second-level sequence. They are not quite so easy to recognize. Look at the following set of numbers and the steps between them.

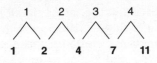

The next step value will be 5, so the next in the series will be 11 + 5 = 16.

Here is another example:

1 Cover the lower lines of this page and work out the differences between the numbers in this series: 44, 56, 71, 89 ...
2 The steps are: 12, 15, 18 ...
3 The operation here is +3, so the next step value will be 21, and the next in the series will be 110.

Here is one more. What's happening here?

1 240, 234, 216, 162 ...
2 The steps are: 6, 18, 54 ...
3 The operation is × 3. The next step value will therefore be 162, and the next in the series will be 0.

Practice 2

Give the next two values.

1) 5, 6, 8, 12, ..., ...
2) 57, 47, 38, 30, ..., ...
3) 0, 30, 90, 180, ..., ...
4) 35.4, 34.8, 33.6, 31.2, ..., ...
5) 500, 257, 176, 149, ..., ...

Dual operation sequences

Where the same operation is performed at each step, a consistent pattern emerges very quickly. If two or more operations are being used, it takes longer to identify what's going on. For example, what's happening here?

0, 1, 2, 3, 6, 7, 14

Look at the steps between the terms and it's clear by the fourth that this is not a simple sequence. Now look at the steps between every other pair of numbers. One sequence is formed by doubling – the 3 to 6 and 7 to 14 pairs stand out. The other is a simple +1.

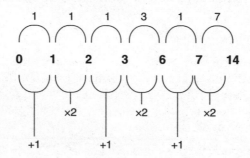

The next two terms would be (+1) 15 and (× 2) 30.

Practice 3

What are the operations in these sequences? What are the next two numbers?

1) 120, 100, 110, 90, 100, 80, 90, ..., ...

2) 0, 0, 2, 6, 8, 24, 26, ..., ...

3) 8, 4, 12, 6, 18, 9, 27, ..., ...

4) 0, 4, 1, 6, 3, 10, 7, 18, ..., ...

Interwoven sequences

These are quite popular with some numeracy testers and can sometimes be quite tricky to spot. They should be visible by the time you reach the fifth or sixth term. Take this sequence. First look at the steps from one to the next:

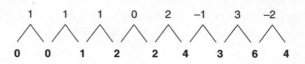

This makes no sense. Try the steps between every other number:

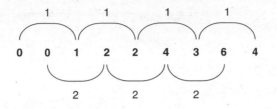

Two simple sequences emerge. The next numbers will be 6 + 2 = 8 and 4 + 1 = 5.

Practice 4

What are the next two numbers?

1) 1, 2, 3, 4, 5, 8, 7, 16, ..., ...
2) 0, 1, 4, 3, 8, 9, 12, 27, ..., ...
3) 96, 69, 82, 78, 68, 87, ..., ...
4) 120, 60, 60, 50, 30, 40, 15, ..., ...
5) 0, 3, 6, 8, 18, 18, 42, 38, ..., ...

Alphabetical sequences

Some employers use letter sequences in their tests, which also help to measure your ability to think logically and analytically. If you do come across sequences of letters in a numeracy test, don't worry. As long as you know your alphabet (!) there's nothing to these – just count the steps between in the same way as for number sequences.

For example:

a, d, g, j, ...

With two steps between each term, this is the same as 1, 4, 7, 10, ... so the next one would be 'm' (13).

You may also find second-level and interwoven sequences using letters, and they may work backwards or forwards through the alphabet. Tackle these as if they had number terms – but note that the alphabet is usually taken as looping round ... x, y, z, a, b, c ...

Practice 5

What comes next?

1) b, f, j, n, ...

2) a, b, d, g, k, ...

3) y, v, s, p, ...

4) z, y, w, s, ...

5) a, b, d, d, g, f, ..., ...

Quirky sequences

You will sometimes find sequences that aren't created arithmetically. For example:

7, 11, 13, 17, 19, ...

The next number would be 23, as these are from the list of prime numbers (those divisible only by themselves and 1). The first 20 prime numbers are:

2, 3, 5, 7, 11, 13, 17, 19, 23, 29, 31, 37, 41, 43, 47, 53, 59, 61, 67, 71.

You might also meet the list of square numbers. Here are the first ten:

1, 4, 9, 16, 25, 36, 49, 64, 81, 100.

Number patterns

You are more likely to meet problems involving number patterns in numerical reasoning tests than in numerical aptitude tests, because solving them relies more on having a 'feel' for numbers than on arithmetical skills. In these kinds of problems, sometimes you need to focus on the digits of a number rather than its value. For instance, you can look at 123 as a value something over 100, but also at the digits 1, 2 and 3.

In the questions that follow, cover the answer below each one and give yourself time to think about the question before you read the explanation.

Q. What do 279 and 628 have in common?

A. You can add the first two digits to get the third.

Q. In the pattern 19: 61 – 68: XX, what is the missing number?

A. 89 – the digits have been turned 180°.

Q. Which is the odd one out of the following: 1,368, 1,863, 3,671, 8,613?

A. 3,671 – the others use the same digits.

Remember these patterns as you do the next test.

Practice 6

1) What's the missing number?

294	722	428
583	790	207
737	?	135

    a) 843   **b)** 736   **c)** 872   **d)** 602

2) What's the missing number?

109	891
168	?

    a) 106   **b)** 601   **c)** 901   **d)** 619

3) What's the missing number?

24	35	48
63	42	54
82	51	?

    a) 66   **b)** 62   **c)** 72   **d)** 60

4) What's the missing number?

5783	5873	3857
3546	3456	6435
5493	?	3954

    a) 5943   **b)** 9543   **c)** 3459   **d)** 5934

5) What's the missing number?

63	20	29
15	10	22
42	26	47

35	16	62
18	?	20
82	22	57

    a) 19   **b)** 11   **c)** 38   **d)** 18

Summary

Today you learned that sequences can be created in several different ways – the trick is to spot which way, as quickly as possible. Tackle them like this:

1 Work out the differences between the terms, writing them down unless they are very easy to remember.

2 If the first three differences are the same, you have a simple sequence and can use the value to calculate the next term.

3 If the differences vary, but are all either increasing or decreasing consistently, work out whether it is a +, −, × or ÷ operation that will turn each difference into the next.

4 If there is no visible pattern to the differences, assume it's an interwoven sequence and look at the differences between alternate pairs.

Where numbers are presented in boxes, circles or other patterns, and you are asked to supply the missing number, look first for arithmetical relationships between the other values. If nothing is visible, look instead at the individual digits that make up the numbers and try manipulating them.

SUNDAY

MONDAY

TUESDAY

WEDNESDAY

THURSDAY

FRIDAY

SATURDAY

End-of-the-day test

[answers at the back]

What comes next?

1) 1, 1, 2, 3, 5, 8, 13, 21 ?
   a) 30
   b) 34
   c) 36
   d) 42

2) 12, 24, 36, 72, 108, 216 ?
   a) 288
   b) 144
   c) 432
   d) 324

3) 11, 12, 14, 19, 20, 23, 28, 29, 33 ?
   a) 34
   b) 39
   c) 38
   d) 40

4) 1, 2, 4, 6, 9, 12, 15, 19, 23, 27, 31 ?
   a) 35
   b) 40
   c) 37
   d) 36

5) 25, 20, 21, 17, 19, 16, 19 ?
   a) 17
   b) 21
   c) 15
   d) 18

6) 3, 4, 5, 8, 12, 17, ...

7) m, n, q, q, u, t, ..., ...

8) 24, 43, 35, 54, 46, ...

9) What is the missing digit?
   8729, 7355, 6?88
   a) 2
   b) 4
   c) 3
   d) 7

10) What's the missing number?

1437	7314	4173
2645	5426	6254
3169	????	1396

   a) 9631
   b) 9361
   c) 1396
   d) 6193

Answers to practice questions

Practice 1:

1) 40 (step = − 4)
2) 75 (step = + 17)
3) 3 (step = ÷ 3)
4) 7.0 (step = + 1.5)
5) 305 (step = + 54)

Practice 2:

1) 20, 36. The differences are increasing by × 2.
2) 23, 17. The differences are decreasing by 1.
3) 300, 450. The differences are increasing by 30.
4) 26.4, 16.8. The differences are 0.6, 1.2, 2.4 – doubling each step.
5) 140, 137. The differences are 243, 81, 27 – divided by 3 each time.

Practice 3:

1) 70, 80 – the operations are − 20 and + 10.
2) 78, 80 – the operations are × 3 and + 2.
3) 13.5, 40.5 – the operations are ÷ 2, add the previous number.
4) 15, 34 – this one was trickier. Write down the steps between the numbers and you get:

+ 4, − 3, + 5, − 3, + 7, − 3, + 11

One operation is clearly − 3, but what is the other? Look at them as a sequence:

4, 5, 7, 11

The steps are + 1, + 2, + 4 – they are doubling each time, so there must be a × 2 in the operation, but that's not all. What else is needed to get from 0 to 4, or from 1 to 6? .... × 2 = 4 – the missing number is 2. Try it on the next pair: (1 + 2) × 2 gives you 6. The operation is + 2 × 2.

Practice 4:

1) 9, 32 – the operation in the first sequence is + 2, in the second it is × 2.

2) 16, 81 – the operations are + 4 and × 3.

3) 54, 96 – the operations are – 14 and + 9.

4) 30, 7.5 – the operations are ÷ 2 and – 10.

5) 90, 78 – the operations are + 3 × 2 and + 1 × 2.

Practice 5:

1) r – simple sequence + 3

2) p – second-level sequence, the steps increase by 1 each time

3) m – stepping back 2 letters at a time

4) k – stepping back, doubling each time.

5) j, h – interwoven sequences, one steps + 2, the other +1

Practice 6:

1) c (add the values from the left and right columns in the row)

2) b (the numbers in the opposite corners are rotated 180°)

3) d (add the digits of the numbers in the first two columns and multiply $(8 + 2) \times (5 + 1) = 60$)

4) a (the order of digits is changed to a set pattern)

5) b (add the digits of the numbers in the left and right columns to get the ones down the middle)

SUNDAY

MONDAY

TUESDAY

WEDNESDAY

FRIDAY

SATURDAY

FRIDAY

Interpreting data from charts, graphs and tables

Data interpretation exercises provide an efficient way to test several key numeracy skills and abilities. You may find the data presented in the form of graphs or charts, or numerically in tables. Either way, to do well in data interpretation problems, you need to be able to:

- know how to get the relevant data
- do arithmetic
- calculate percentages
- use estimating techniques to solve problems quickly.

In fact, you could use pretty well everything that we have done so far this week in a data interpretation problem.

One last point: with these tests, sometimes you will be asked a question when the data you need in order to give an answer simply isn't present. It's not a trick but a way to check that you can identify relevant data. Therefore watch out for 'Can't tell' options in multiple-choice questions. They are there for a reason.

Bar charts

If you want to compare multiple sets of data, such as sales of different products over time or production of goods in different countries, bar charts – or column charts, as Excel calls them – offer an effective way to display them.

A bar chart will take its data from a table like this:

Sales by store

	Spring	Summer	Autumn	Winter
London	135	186	194	164
Bristol	85	126	134	118
Cardiff	150	174	182	143

And will look something like this:

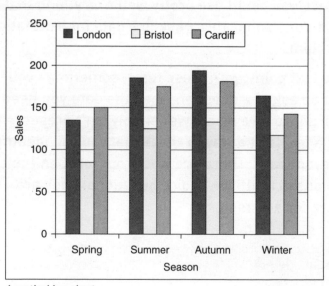

A vertical bar chart

Each row of figures becomes a set of vertical bars of the same colour, and a legend – usually to the side – links these colours to the row headings. (If the colours are not distinctive enough, you can identify the bars by their position. The first one in each group is the first item in the legend, the second is next and so on.) The height of the bars is determined by the values, and the scale of values is shown in the vertical (y) axis. The column headings are written across the horizontal (x) axis – along the bottom of the chart.

Although there are variations, all bar charts follow the same general rules. The bars can be drawn horizontally, like this:

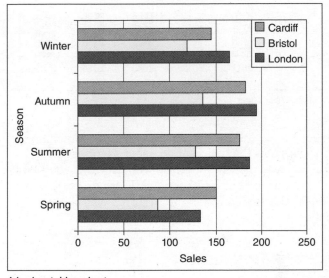

A horizontal bar chart

Here the x axis is now vertical and the y axis is horizontal, but you still read the data in the same way. If you wanted to know, say, the sales for the Cardiff office in summer, you would first locate summer on the x axis, then use the legend to identify the Cardiff bar, and then draw a (mental or pencil) line from the end of the Cardiff bar across or down to the y axis. It crosses the axis mid-way between 150 and 200, giving a value of 175.

You also get stacked bar charts (which can be either vertical or horizontal), where the values at each point on the x axis are stacked on top of one another instead of being side by side. It's easy to read total values from these, but much harder to compare the sets of data.

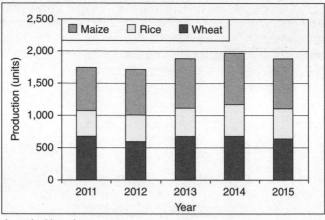

A stacked bar chart

Questions on totals are simple. For example, a question asking which years saw the highest and second highest levels of production would be simple to answer. However, getting data out of the stacked bars is harder work; for example a question asking for the year in which the rice crop was poorest would be a challenge.

Accuracy and answers

The values you get from charts are rarely exact, and they are not meant to be. Charts are there to allow you to make rough comparisons and to see overall trends and changes. Chart-based questions are normally multiple choice, and your inexact calculations should get you close enough to identify the correct answer.

Under test pressure, it's all too easy to make mistakes. Always double-check the question and the legend to make sure you are looking at the right bars.

Practice 1

Allow yourself five minutes for each of the exercises today.

The chart shows world cereal production, in 000s of tonnes, from 2011 to 2015. Answer the following questions, to the nearest 10,000 tonnes.

1] How much larger was the maize crop than the wheat crop in 2014?

   a] 350,000   b] 130,000   c] 170,000   d] 680,000

2] What was the difference between the maize and rice crops in 2015, expressed as a percentage of the rice crop value?

   a] 38%   b] 152%   c] 77%   d] 85%

3] In which year was maize not the largest crop?

   a] 2011   b] 2013   c] 2014   d] Never

4] What was the total rice crop in the years 2011 to 2015?

   a] 490,000 tonnes   b] 2,450,000 tonnes
   c] 1,870,000 tonnes   d] 2,150,00 tonnes

5] Which year saw the biggest increase in wheat production?

   a] 2011   b] 2013   c] 2014   d] 2015

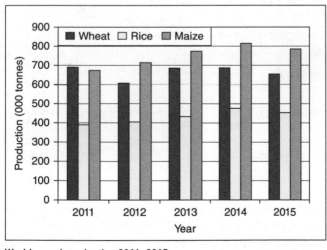

World cereal production 2011–2015

Line graphs

Line graphs are good for showing trends and can be used for finding intermediate values. Take this table of data, showing the populations of France, Italy and the UK between 2005 and 2013.

	2005	2007	2009	2011	2013
Italy	58,103,030	58,147,730	58,126,210	60,316,800	60,782,670
France	60,656,180	63,713,930	64,057,790	65,312,250	65,856,610
UK	60,441,460	60,776,240	61,113,200	62,698,360	64,308,261

The data can be represented as a line graph like this:

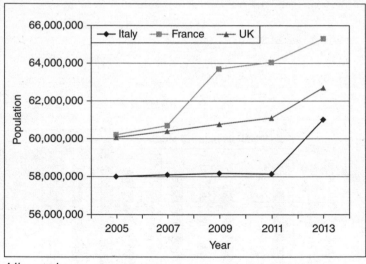

A line graph

It's essentially the same structure as a bar chart. Each row of figures is plotted here as a line, identified by colour or style or markers, and a legend links the lines to the row headings. The height of the points on each line is determined by the values, and the scale of values is shown in the vertical (y) axis. The column headings are written across the bottom of the chart, along the horizontal (x) axis.

Questions based on this graph might ask:

1 When did the population of Italy reach 60 million?
If you drop a line down from the point where the Italy line crosses the 60 million horizontal, it will fall in late 2010. You can do this by eye or with a pencil (drawn on paper or held up to the screen).

2 When was the difference between the populations of France and Italy greatest?
It looks like 2009, but check with a ruler or the edge of a piece of paper marked with the top and bottom points.

3 If the trend from 2011 to 2013 continues, which country will have the lowest population by 2017?
Italy is some way behind and can be ignored. Projecting the France and UK lines forward will see them cross just off the edge of the graph, putting the UK at the highest by 2015 or 2016.

4 What was the population of France in 2006?
Draw a line up from the marker between 2005 and 2007 to cut the France line. Draw across from this to the *y* axis, which it cuts at something over 60 million.

Practice 2

The following graph shows the driving test pass rates of two schools of motoring and the national average.

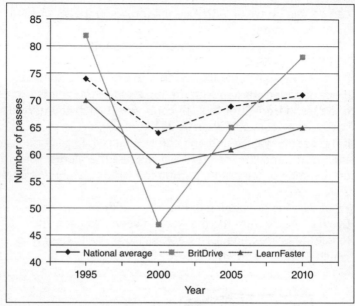

Percentage driving test pass rates, 1995–2010

1) In which year was the LearnFaster pass rate most below the national average?

   a) 2000   b) 2005   c) 2010   d) Can't tell

2) For how many years was BritDrive's pass rate below the national average?

   a) 6   b) 8   c) 10   d) 12

3) BritDrive put 640 candidates in for the test in 2010. How many failed?

   a) 499   b) 115   c) 126   d) 141

4) When did LearnFaster's pass rate get back above 60% again?

   a) 2002   b) 2004   c) 2006   d) Can't tell

5) There were 24,500 candidates in 2010. How many passed?

   a) 17,395   b) 18,643   c) 18,667   d) 19,547

Pie charts

Pie charts are used to show proportion – how a whole is shared out among its parts. The techniques that you use with fractions, percentages and ratios come into play with these charts.

Pie charts normally only display the values from a single data series – one row in a table. For example, here is the average monthly expenditure for a household:

Expenditure, 2010	Average per month, £
Rent/mortgage	650
Power	120
Food	440
Transport	180
Clothes	45
Entertainment	80
Savings	50

Presented as a pie chart, it might look like this:

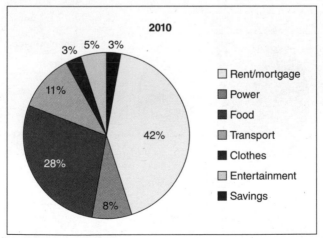

A pie chart, 2010 household expenditure

Slices can be identified by their order if the colouring is not distinctive enough. Starting from the top and going round clockwise, the slices are in the order that they are listed in the legend. The values here have been changed to percentages but sometimes they are left at the actual values. Numeracy tests tend to use percentages because then they can ask questions like these:

Q. If the total monthly expenditure is £1,565, how much is spent on transport?

A. You find this by reading 11% off the chart, then calculating 11% of £1,565.

Q. If the cost of power goes up by 20% and all other costs remain the same, how much would be left for saving?

A. Power is 8%. 20% of 8% = 1.6%. Saving is currently 3%, and would therefore reduce to 1.4%.

Sometimes you will see two charts side by side, showing the same categories of data but taken from different years, countries or whatever.

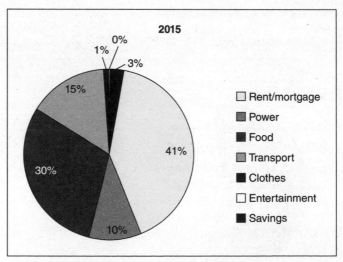

A pie chart, 2015 household expenditure

For this example, you might then be asked questions like the following.

Q. Comparing power and food costs in 2010 and 2015, which showed the greatest proportional increase?

a) Power

b) Food

c) Both the same

A. a. Power has gone up from 8% to 10%, and food up from 28% to 30%. Both are 2%, but as a proportion 2/8 (=1/4) is much higher than 2/28 (=1/14).

Q. If the total monthly expenditure in 2010 was £1,565 and in 2015 it was £1,600, by how much has the cost of food increased?

A. In 2010, food was 28% of £1,565 = £438. In 2015 it was 30% of £1,600 = £480. The increase is £480 – £438 = £42.

Data from tables

When data is presented graphically, the key skills are picking out the relevant values and using estimates to identify correct answers. When the data is given in the raw – as lists and tables of numbers – it becomes even more important to be able to pick out the relevant values. A table may well have half a dozen or more rows and columns, with 40 or more values, of which only one or two may be relevant. To succeed in these kinds of questions, you should take the following steps:

1 Read the question carefully.
2 Use the row and column headers to identify the data you need.
3 Read the question again to check that it is the right data.
4 Do whatever calculations are needed – and typically they involve totals and percentages.

For example:

Q. A survey of how children travel to school produced the following data.

Age	Walk	Cycle	Scooter	Car	Bus
4–5	32	2	6	14	4
6–7	33	6	4	12	6
8–9	28	13	5	11	5
10–11	26	17	4	6	7

1 What percentage of 4–5 year olds walk to school?
2 What is the total number of 8–9 year olds?
3 How many children under 8 years old travel by cycle or scooter?
4 What is the ratio of car travel to all other forms of travel in the 10–11 year group?

Answers

1 55%. The sum is *Walk/Total* as a percentage, so find the total of the group (58), then divide *Walk* by that (32 ÷ 58 = 0.5517) and convert to a percentage.
2 62: simple addition.
3 18: add the values in the *Cycle* and *Scooter* columns of the 4–5 and 6–7 rows.
4 1:9. There are 54 others in total, so the ratio is 6:54, which simplifies to 1:9.

Practice 3

This pie chart shows the proportion of seats held by the different parties in the UK House of Commons in 2015.

1) There were 650 seats in total. How many were Labour?

    a) 192   **b)** 219   **c)** 232   **d)** 301

2) How many more seats did the Conservatives have than all the other parties together?

    a) 10   **b)** 18   **c)** 22   **d)** 30

3) How many seats did the small parties (from Liberal Democrat down) have in total?

    a) 14   **b)** 27   **c)** 33   **d)** 42

4) Before the 2015 election, the Conservative party had 305 seats. By what percentage did their number of seats increase?

    a) 5%   **b)** 9%   **c)** 15%   **d)** 25%

5) If, after the next election, the Conservatives lose 15% of their seats to the Labour Party and all other seats are unchanged, could they get a majority by forming a coalition with the Liberal Democrats and the other smaller parties?

    a) Yes   **b)** No   **c)** Too close to call

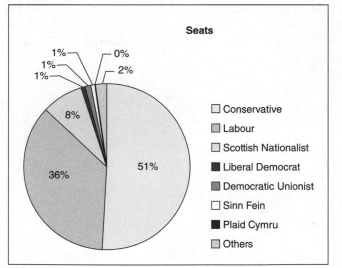

A pie chart, House of Commons seats, 2015

Summary

Today you learned that your ability to interpret data reflects several key numeracy skills and abilities. In a numeracy test you may find this data presented in the form of graphs or charts, or numerically in tables.

- The values you get from charts are rarely exact, so chart-based questions are usually multiple choice, and you can normally use approximate values to identify the correct answer.
- Bar charts can be drawn with the bars vertical or horizontal. Whichever way round they are, the x axis shows the column headings from the data table behind the chart and the y axis shows the scale – value, volume, quantity or other variable. The bars are identified by colour and by their order in the legend.
- Line graphs have the same basic structure as bar charts and can be used to represent the same data sets. They are good for showing trends and for finding intermediate values.
- Pie charts normally only display the values from a single data series – one row in a table. The values are often given as percentages.

SUNDAY
MONDAY
TUESDAY
WEDNESDAY
THURSDAY
FRIDAY
SATURDAY

End-of-the-day test

[answers at the back]

Bar chart

This chart shows bicycle sales by four local retailers over a financial year.

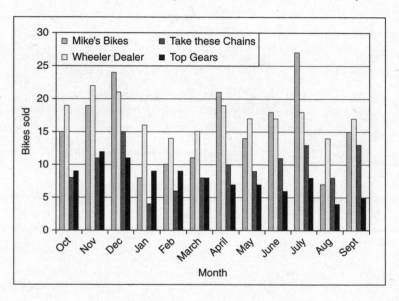

1) Which retailer has the most consistent sales over the year?

2) In which month were most bikes sold by the four combined?

3) In which month did *Mike's Bikes* and *Take these Chains* sales differ most?

4) What was the percentage increase in *Wheeler Dealer* sales between August and September?

5) Would you expect *Top Gears* sales next year to be about the same, more or less?

Line graph

The graph shows exchange rates for the euro, dollar and Swiss franc against the GB pound from January 2008 to January 2012.

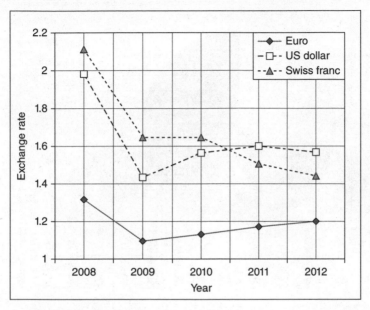

6) When the Swiss franc and the US dollar were the same value, how many could you get to the pound?

   a) 0.66   b) 1.62   c) 1.56   d) 1.45

7) If I converted €500 to pounds in January 2008, then back to euros in January 2012, how many would I have?

   a) €540   b) €453   c) €320   d) €500

8) When is the greatest difference between the US dollar and Swiss franc rates?

   a) 2008   b) 2009   c) 2011   d) 2012

9) If I converted $1,000 to Swiss francs in January 2008, then back to US dollars in January 2012, how many would I have?

   a) $860   b) $1,000   c) $1,050   d) $1,160

10) By what percentage did the GB pound/US dollar exchange rate change from 2008 to 2009?

   a) 15%   b) 21%   c) 27%   d) 38%

Pie chart

This chart shows the sales, by size, of pizzas in a restaurant during one evening when a total of 160 pizzas were sold.

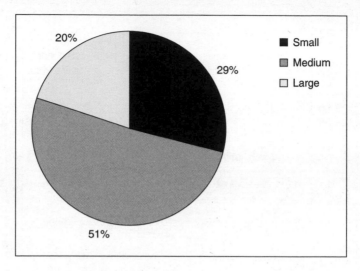

11) How many small pizzas were sold?

a) 29   b) 35   c) 42   d) 47

12) If sales of small pizzas doubled the next day, but medium and large sales stay the same, what percentage of sales would be small?

a) 58%   b) 45%   c) 41%   d) 29%

13) Pizzas cost £5.00 for small; £8.00 for medium and £11.00 for large. What is the total value of sales?

a) £1,235   b) £1,057   c) £965   d) £773

14) What percentage of the total sales income came from large pizzas?

a) 20%   b) 23%   c) 29%   d) 38%

Data from tables

The following table is drawn from census data for a small town.

Year	1999	2004	2009
Total	3,357	3,263	3,189
Male	46%	45%	43%
Female	54%	55%	57%
Child/student	20%	20%	18%
Retired	30%	36%	38%

15) In 2009, how many males were there?

16) What is the percentage change in the number of females between 1999 and 2009?

17) How many people were of working age in 2004?

18) The 2004 unemployment rate was 17% (of working population). How many people had a job?

19) In 2009, how many women were retired?

Answers to practice questions

Practice 1:

1) b (maize = 810; wheat = 680; 810 – 680 = 130 = 130,000 tonnes)

2) c (maize 780; rice 440; difference = 340; 340/440 × 100 = 77%)

3) a (the wheat bar is fractionally higher in 2011)

4) d (390 + 400 + 430 + 480 + 450 = 2,150 = 2,150,000)

5) b.

Practice 2:

1) b. The average rate increases faster than LearnFaster from 2000 to 2005, widening the gap, and then its rate of increase slows down and LearnFaster catches up.

2) c. BritDrive's line crosses the average on the way down about 2/5 of the way between 1995 and 2000 (i.e. 1997), then crosses again on the way up 2/5 of the way between 2005 and 2010 (i.e. 2007).

3) d. In 2010 BriDrive's pass rate is something under 80%, so at least 20% failed. 20% of 640 = 128 and the actual number must be more than this.

4) b. It's either 2003 or 2004 and only 2004 is available as a choice.

5) a. 70% of 24,500 = 17,150 and the actual rate is higher – but only slightly.

Practice 3:

1) a. Labour has 36%. Round up and find 40% of 650 = 260. The answer will be something less than this. With a calculator, 36% of 650 = 234. The percentages have all been rounded, so 36% is not exact. Take the closest answer.

2) b. The Conservatives have 51%, leaving 49%. The difference is 2%. 2% of 650 = 13. There will have been some rounding – either up or down, or both – but 10 is the closest answer.

3) c. The slices add to 5%. Rounding errors may have reduced this, so check the total of the big three. That's 95%. 5% of 650 = 32.5. Therefore 33 is the closest.

4) d. The Conservatives have 51%. 50% of 650 = 325. The increase is 325 – 305 = 20. Round to 20/300 = 7%. 8% is closest. With a calculator, 51% of 650 = 331.5 – 305 = 26.5/305 = 8.6%

5) b. You can do this in percentages, rather than work out the number of seats. Round the Conservative figure to 50%: 15% of 50 = 7.5%. 51% – 7.5% = 43.5%. The Liberal Democrats and smaller parties have 5% between them, which takes the total up to 48.5%. There could be some rounding errors here, but these are unlikely to be enough to make up the extra percentage to give them a majority.

SATURDAY

Test techniques and practice

This is the final day of your week's work, and you are almost there. By now you should have all the numeracy skills you need in order to achieve mostly correct answers in a numeracy or numerical reasoning test. Today the focus is on learning techniques that will maximize your test scores. These techniques are:

- getting into the right frame of mind
- using the time well
- avoiding howlers
- choosing wisely from multiple choices.

Since the most important part of your preparation for a numeracy test is the daily practice of the skills you will need, we finish the day with two full-length tests, one based on calculation, the other on data interpretation.

Test techniques

There are various techniques to remember once you are in the test situation. Knowing these can make the difference between success and failure, even when you are feeling confident about your numerical ability.

You can do it – mostly...

If a numeracy test is part of a selection process, it is there to help employers discriminate between candidates. If the test is so easy that everyone can get 100% or so hard that no one can do it, it is useless. If you are a suitable candidate for the job, you should be able to answer most of the questions but not necessarily all of them.

Like many people, you may be reminded of your school experience and feel fear at the very thought of doing maths-related tasks under timed conditions. However, most of these tests are easier than they seem at first, especially if you have reviewed and understood the concepts outlined in this book and used the tests to practise the skills diligently. The key is to go at it confidently, but without being thrown if you come across a question that you can't do.

Stay calm when faced with a question you think you can't do: remember that you won't always need to be able to answer all the questions and, in any case, it may be easier than you first think.

Timing

These tests are always against the clock, but the testers will be allowing you enough time to show your capabilities. So don't panic, but do use the time wisely. The test may advise you how long to spend on each question but, if it doesn't, you need to work this out.

If it is on paper or on screen in a form that allows you to go back and forth through the questions, take a few moments at the start to skim through to get the sense of what's there. If there are similar questions throughout, divide the total time by the number to give you the amount of time you will have to spend on each one. If the questions get harder, note the one that comes about two-thirds of the way through and aim to get there by halfway through the test time.

Keep an eye on the clock as you work. If you find you are getting bogged down in a question, leave it and move on. If time allows, you can come back to it later.

Checking

Always take a moment to check your answer. This is just as important with easy questions as with harder ones – possibly more so, as we tend to give them less time and less attention. Here are some ways to avoid making silly mistakes.

- Use nice round numbers to check the size.
- If you can do so easily, check the final digit.
- If appropriate, check that you have used the right units of measurement – km, metre, cm, mm, $, £, p, etc.

Multiple-choice techniques

The key thing to bear in mind with multiple-choice questions is that you do not have to work out the answer. Instead, you have to work out which of the offered answers is the right one, and that presents a completely different situation.

With multiple choice, one of the options given is the right answer. If you cannot work it out, or have run out of time, take a guess – that will give you a 1 in 4 (or 1 in 3) chance of a score. If you leave it blank, you score nothing.

However, sometimes one answer will be very different from the rest. It's there to catch people who guess without thinking. If this is the case, use a rough estimate to filter out the answers that are outside the possible range. This will improve your chances if you guess.

Sometimes it is simpler to work backwards from the possible answers. For instance, if you have to do a division to solve the problem, you can multiply back from the possible answers. If you don't have a calculator, it may be quicker and more reliable to do two or three multiplications than one division.

Summary

Today you learned some test techniques to raise your performance in a test situation. However, there is no substitute for sound knowledge of numeracy skills and knowing how to apply them. Revisit the chapters in this book and focus especially on the areas about which you are less confident. Make sure you understand each section by going back over the practice tests and the questions at the end of each day and analysing the reasons for your errors.

Practice may not make you perfect, but it will certainly improve your performance. Try the 30-minute tests that follow here under timed conditions. There are also some good sources of practice tests online and these are listed at the end of the book. Some of these are free; others are charged for, but the rates are generally reasonable and could be a good investment if they help get you that job.

Good luck!

SUNDAY

MONDAY

TUESDAY

WEDNESDAY

THURSDAY

FRIDAY

SATURDAY

Test 1

Allow 30 minutes to complete the test. You may use a calculator.

1) What comes next: 1, 1, 7, 4, 13, 7, 19, 10, ...?

    a) 23   **b)** 25   **c)** 16   **d)** 29

2) What comes next: 100, 86.6, 73.2, 59.8, 46.4, ...?

    a) 34.6   **b)** 35.2   **c)** 32.8   **d)** 33

3) What is the missing number?

5	8	6	7
3	2	6	8
4	4	5	6
2	3	3	?

    a) 2   **b)** 9   **c)** 7   **d)** 4

4) Which block will fill the space?

7	4	11	15
11			35
9		16	23

a)		b)		c)		d)	
19	27	8	13	12	23	9	18
7		6		7		13	

5) What is the missing number? 4672   3496   ?595

    a) 2   **b)** 4   **c)** 3   **d)** 7

6) What is the missing number?

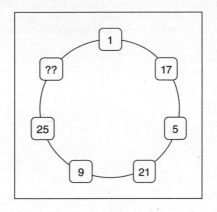

a) 21   b) 34   c) 13   d) 7

7) What is the missing number in the third box?

	6			7			8	
5	37	7	8	62	6	3		5

a) 29   b) 77   c) 53   d) 437

8) What comes next? 1, 2, 5, 14, 41, ...
   a) 114   b) 95   c) 55   d) 122

9) A train travelling at 90 kph enters a tunnel 10 km long. The train is 0.5 km long. How many minutes does it take for the entire train to pass through the tunnel?

10) It's half a mile from Naomi's flat to the gates of the park, three-quarters of a mile across to the gates on the opposite side, five-eighths of a mile from there down to the canal, and half a mile back along the canal home. How far is her morning jog?

11) At the village bring-and-buy sale, Marge and Flora were both selling cake. The cakes were the same size and thickness. Marge cut hers into 6, and then cut each of these pieces into 4. Flora cut hers in half, then cut each half into 5. Marge was offering 2 pieces for 50p; Flora offered 1 piece for 50p. Which was best value?

12) When Uncle Ebeneezer died, he left £150,000 to be divided this way: one third to his faithful housekeeper of six months, Fifi LaTouche; one sixth to his only son, and the remainder to be divided equally between his five daughters. How much did each daughter get? What fraction was this?

13) In our industry, turnover has fallen in the recession by an average of 8%. Our turnover dropped from £875,050 to £794,000. Are we doing better or worse than average?

14) On an assembly line, they aim for a maximum of 1.5% faulty units. If 7,450 units are produced in a day and 134 are faulty, is this within the acceptable limits?

15) Jake worked these hours last week: Mon 9.5, Tue 10.0, Wed 8.75, Thurs 9.0, Fri 8.5. What was the daily average?

Test 2

Allow 30 minutes to complete the test. No calculators this time!

A. Mobile tariffs

The Talkathon mobile phone company offers the following tariffs.

- Freedom: £40 per month, unlimited calls and texts
- Standard: £20 per month, 100 minutes of calls, 100 texts. Additional calls 9p per minute, texts 5p each.
- Pay As You Go: Calls 15p per minute, texts 10p each.

1) Sally makes an average of 240 minutes of calls and 150 texts a month. Which is the best tariff for her?

2) George makes an average of 80 minutes of calls and 70 texts a month. Which is the best tariff for him?

3) If a person sends 250 texts a month, how many minutes of calls do they need to make for Freedom to be the best tariff?

B. Smartphone sales

The first of the following graphs shows the sales of smartphones in 2014, by gender, in thousands. The second shows the breakdown of all smartphone buyers by age group.

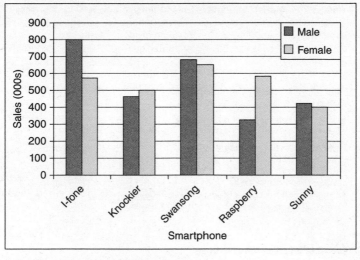

Sales of smartphones in 2014, by gender

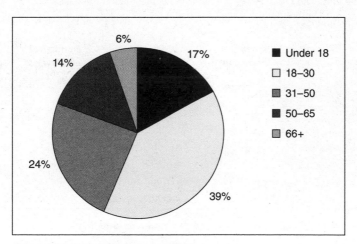

Smartphone buyers by age group, 2014

1) Roughly what percentage of Raspberry phones were bought by males?
    a) 28%   b) 36%   c) 42%   d) Can't tell

2) Which company sold the most phones in total?
    a) I-fone   b) Swansong   c) Raspberry   d) Can't tell

3) What is the approximate male to female ratio among I-fone buyers?
   a) 2: 1
   b) 1: 1.1
   c) 1:1.4
   d) 1:1.7

4) There was a total of 540,000 buyers. How many were in the 31–50 age group?
   a) 1,207   b) 1,300   c) 1,456   d) Can't tell

5) How many female Sunny buyers were in the 18–30 age group?
   a) 1,045   b) 955   c) 1,125   d) Can't tell

6) Knockier sales to female buyers fell by 20% in 2015. If they fell another 20% in 2016, how many buyers would there be?
   a) 250,000   b) 300,000   c) 320,000   d) 400,000

C Assembly line machine failures

The following table shows the average weekly downtime caused by machine failure on four assembly lines.

Line	Average downtime (h)	Cost of lost revenue	Number of employees
A	8.5	€20,995	8
B	7.25	€5,474	2
C	5.75	€17,768	14
D	2.25	€5,018	12

1) Which line loses most revenue per hour due to machine failure?
   a) Line A   b) Line B   c) Line C   d) Line D

2) Which line loses least revenue per employee due to machine failure?
   a) Line A   b) Line B   c) Line C   d Line D

3) If machine failure could be reduced by 15% on each line, how much revenue would be gained each week?
   a) €4,675   b) €6,475   c) €7,388   d) Can't tell

4) How many employee hours are lost in total due to machine failure?
   a) 36   b) 148   c) 178   d) 190

Test 1 answers and explanations

1) b (two intermixed series, adding 6 and 3)

2) d (subtract 13.4 each time)

3) a (the columns total 14, 17, 20 – next must be 23)

4) c (add columns 1 & 2 to get column 3; add columns 2 & 3 to get column 4)

5) b (multiply the middle numbers to get the one shown in the outside digits)

6) c (starting from 1, jump a number and add 4 to the next)

7) a (multiply top and left numbers, then add the one on the right)

8) d (multiply by 3 and subtract 1 to get the next)

9) 7 minutes. The total distance travelled is 10.5 km, so the sum is $10.5 \div 90 \times 60$.

10) The sum of Naomi's run is: $1/2 + 3/4 + 5/8 + \frac{1}{2}$. We can simplify this to: $1/2 + 1/2 = 1 + 3/4 + 5/8$. The common denominator is 8: $1 + 6/8 + 5/8 = 1 + 11/8 = 1 + 1 + 3/8 = 2\ 3/8$

11) Marge cut her cake into $6 \times 4 = 24$ pieces, but sold them 2 at a time. $2/24 = 1/12$. Flora cut hers into $2 \times 5 = 10$. Since $1/10$ is more than $1/12$, Flora's cake was best value.

12) There are several sums here:

    $1/3$ of £150,000 = £50,000 for Fifi LaTouche

    $1/6$ of £150,000 = £25,000 for the son

    That leaves: £150,000 – £50,000 – £25,000 = £75,000.

    As a fraction: $1/3 + 1/6 = 2/6 + 1/6 = 3/6 = 1/2$ (to Fifi and the son)

    $1 – 1/2 = 1/2$ the fraction left

    £75,000 $\div$ 5 = £15,000 for each daughter. As a fraction: $1/2 \div 5 = 1/10$.

13) A fall of £875,050 to £794,000 is £81,050. This is a bit less than 10% of £875,050. As a rough check, we are probably doing worse than average. Accurately:

    $81,050 \div 875,050 \times 100 = 9.26\%$

14) 134 out of 7,450 = As a rough check, 1% of 7,000 = 70; 0.5% = half of that = 35, so 1.5% = something over 105. 134 is therefore probably too high. Accurately: $134 \div 7,450 \times 100 = 1.79\%$.

15) Add up the numbers and divide by 5 to get the daily average, which is 9.15.

Test 2 answers and explanations

A. Mobile tariffs

1) Standard. £20 basic + 140 minutes at 9p = £12.60 plus 50 texts at 5p = £2.50, so a total of £35.10 a month.
Pay As You Go would cost 240 × 15p = £36 + 150 × 10 = £15. Total £51.

2) Standard. The Pay As You Go cost would be 80 × 15 = £12 + 70 × 10 = £9. Total £21.

3) On Standard the extra texts would cost 160 × 5p = £8.00. Add to the base £20 and this is £12.00 short of the Freedom tariff. £12.00 would buy 1,200 ÷ 9 = 133 minutes.

B. Smartphone sales

1) b

2) a

3) c

4) b

5) d. If the age breakdown was the same for every phone, the answer would be 1,045, but we do not know those figures.

6) c. Taking the 2014 figure as 500, reduce this by 20% = 400, then reduce that by 20% to get 320. Multiply by the thousand to get 320,000. (Adding the two 20% reductions together – and knocking off 40% – would give you 300, which is wrong.)

C. Assembly line machine failures

1) Line C. Divide the lost revenue by the number of hours. Line B is clearly much less than A and can be ignored.

2) Line B. A is €2,624, B is €2,737. A rough estimate shows C is a bit over €1,000 and D a fraction of that.

3) c. The total loss is €49,254 × 15% = €7,388.

4) d. Work out lost hours times employees for each line, and add.

7 × 7

1 Seven key ideas

- Never rely on a calculator alone – it's very easy to mis-key. Always check the answer with an estimate.
- It may be quicker to do things in your head, but for anything other than the simplest calculations and problems, write them down on paper. If you can see your workings, you can spot mistakes.
- When working with fractions, it helps if you can visualize them – in your head or sketched on paper.
- When doing conversions, you need to know the conversion rate and which unit is the larger – this tells you whether to multiply or divide.
- Some problems appear to be more complicated than they are. If you don't have the numbers to solve what you think is the problem, try from the other end. Start from the numbers and ask yourself what problem they could solve.
- When data is presented graphically, the key skills are picking out the relevant values and using estimates to identify correct answers.
- Don't panic if you meet a problem that you can't do. It may be too difficult, or you may be looking at it the wrong way. Move on, and come back to it later.

2 Seven of the best resources

- *Great at My Job, but Crap at Numbers* by Mac Bride and Heidi Smith (Teach Yourself) is for you If you've forgotten the maths you learned at school and need a non-patronizing revision course. See also its website: www.crapatnumbers.net
- **SHL** are recruitment specialists whose assessment tests are widely used by employers. You can find practice tests for numerical reasoning, along with those for verbal reasoning, judgement, reading comprehension and personality assessments at their SHL Direct site: http://www.shldirect.com/
- **Assessment Day** offers free practice tests and paid-for banks of test papers across a wide range of aptitude tests, including numerical reasoning (data interpretation). Find them at: http://www.assessmentday.co.uk/
- **Job Test Prep** also offers free practice tests and sells packs of tests for numerical reasoning and other assessments. They are at: http://www.jobtestprep.co.uk/
- **Practice Aptitude Tests** offers numerical reasoning test for free. They are designed for the more challenging end of the jobs market: http://www.practiceaptitudetests.com/numerical-reasoning-tests/
- **Aptitude test** is a good source of free aptitude tests, numeric and other: http://www.aptitude-test.com
- **Graduates First** offers practice tests: http://www.graduatesfirst.com

3 Seven tricks with numbers

- Work on your estimating skills. Check your calculations by doing them again with nice round numbers. If the answer is very different, you have almost certainly made a mistake.
- Where a sum has a mixture of operations, work out the multiplication and division ones first, then the addition and subtraction. If there are brackets, do the operations in these first.
- When working with decimals, the position of the decimal point is the single most important thing to note.
- It's useful to be able to switch easily between fractions, decimals and percentages. Learn the key equivalents – ½, ¼, ¾, ⅕, ⅒.
- When working with fractions, it helps to visualize them. If you have a feel for 'how much of a whole', you can use that to check that the answer makes sense.
- When the data is given in a table, there will be many items of which only one or two may be relevant. To succeed with these, read the question carefully and use the row and column headers to identify the data you need.
- Number pattern problems can be based on the order or appearance of the digits, as well as on the value of the numbers.

4 Seven great quotes

- 'There are no secrets to success. It is the result of preparation, hard work learning from failure.' General Colin Powell, US Secretary of State, 2001–5
- 'It's not that I'm so smart, it's just that I stay with problems longer.' Albert Einstein, physicist and philosopher
- 'The best way to finish an unpleasant task is to get started.' Anonymous
- 'Success seems to be connected with action. Successful people keep moving. They make mistakes, but they don't quit.' Conrad Hilton, founder of the Hilton Hotels chain
- 'When I was young I observed that nine out of ten things I did were failures, so I did ten times more work.' George Bernard Shaw, playwright
- 'Well, tests ain't fair. Those that study have an unfair advantage. It's always been that way.' Allan Dare Pearce, *Paris in April*
- 'When it comes to understanding numbers there are three types of people: those who can and those who can't.' Anonymous

5 Seven things to keep in mind

- If it says 'average' it probably means the arithmetical mean – what you get by adding all the values and dividing by how many there are.
- When a problem is presented in words, not just numbers, you need to make sure that you get from it the sum, the whole sum and nothing but the sum.

- Problems about area, distance, position and similar can usually be solved more easily with the help of a diagram.
- Graphs and charts questions are about making comparisons and spotting trends. Don't fret about getting exact values out of the diagram – near enough is almost always good enough.
- If a test is so hard that you can hardly answer any questions – and the test truly reflects the nature of the job – then you probably wouldn't enjoy the job if you got it.
- But if it's so easy that you can answer them all correctly with time to spare, you may find that the job offers you very few challenges and little satisfaction.
- You won't know what you can do until you try.

6 Seven test tips

- Prepare yourself for the specific test. Find out how and where it will be administered – online, individually, at an assessment centre, etc.
- Take practice tests until you feel fully prepared to take the real tests.
- Get a good night's sleep, and don't eat heavy foods before the test.
- At the start of the test, divide the total time available by the number of questions to see how long you can spend on each – but try to do the earlier ones faster, so you have more time for the later ones which will probably be harder.
- Read questions carefully and check your answers.

- If you are stumped on a multiple-choice question, eliminate any obviously wrong answers if you can, then guess. It's better than leaving it blank!
- If you don't know how to tackle a question, move on. Return to it later if there is time.

7 Seven things to do today

- Practise your tables – remembering that the only ones you really need to learn are 2, 3 and 7.
- Practise using the complements – the pairs of numbers that add up to 10 – on (fairly) simple additions and subtractions, so that using them becomes second nature.
- Practise converting any large or very small numbers that you see around you into nice round numbers, so that you can do it quickly.
- Practise using your calculator, to make your typing faster and more accurate.
- Practise reading text-based problems. Look again at one that you skipped or got wrong on a previous test. Work back from the correct answer, using the explanation – if given – to try to understand what the problem was really about and where its numbers came from.
- Practise interpreting the graphs and charts that you see in newspapers and on websites. What do they tell you? What comparisons could be made?
- Do one of the tests in this book, or one from a website. Practice may not make perfect but it certainly helps!

End-of-the-day test answers

Sunday:

1) 162	2) 197	3) 313	4) 798
5) 552	6) 1,107	7) 14	8) 36
9) 4	10) 420	11) b	12) a
13) b	14) c	15) d	16) b
17) d	18) c	19) a	20) d
21) a	22) b	23) b	24) a
25) b	26) b	27) a	28) b
29) a	30) b		

Monday:

1) 27	2) 160	3) 209	4) b
5) c	6) a	7) c	8) d
9) d	10) b	11) c	12) b
13) 376 ¾	14) 1/10	15) 73 23/25 (73.92)	16) 615.7
17) 7.0002598	18) 95.6	19) 30.555	20) 6.0
21) True	22) False	23) False	24) True
25) True	26) False	27) True	28) False
29) True	30) False		

Tuesday: 1) c; 2) a; 3) b; 4) b; 5) c; 6) a; 7) d; 8) c; 9) a; 10) d;

11) £873.75 (rough check: 250% (2.5 times) 300 = 750, and it's a good bit more than 300)

12) $13.32 (rough check: 25 × 4 = 100 and 15% of $100 = $15)

13) £2.28 (rough check: 6 + 8 + 4 + 1 = 19 × 10% = £1.90)

14) 14 : 28 : 56 (the ratio has 7 parts in total. 98 ÷ 7 = 14)

15) £668,359.08 : £445,572.72 : £445,572.72 : £891,145.44

16) 4.445 m (14 × 12 + 7 = 175 in × 2.54 = 444.5 cm = 4.445 m)

17) €392.40. (Dollars are worth less than pounds so multiply by the fraction to get a smaller number: 500 × 0.62 = 310. Pounds are worth more than euros, so divide by the fraction to get a bigger number: 310 ÷ 0.79 = 392.40.)

18) 144 (2.4 m = 240 cm ÷ 2.54 = 94.5 in ÷ 6 = 15.7 tile lengths. 1.3 m = 130 cm ÷ 2.54 = 51.2 in ÷ 6 = 8.5 tile lengths. In whole tiles: 16 × 9 = 144 tiles)

19) £22,500. (Find the total salary bill: 100,000 + 2 × 40,000 + 25 × 18,000 = £630,000 and divide by the number of staff = 28.)

20) 5. (Multiply the number by the size in each column, then add: 4 × 3 = 12 + 8 × 4 = 32 + 12 × 5 = 60 + 7 × 6 = 42 + 2 × 7 = 14. Total = 160. Add the number values to find how many children: 4 + 8 + 12 + 7 + 2 = 33. Divide the total by the number of children = 160 ÷ 33 = 4 remainder 28. Nearest shoe size is 5.)

Wednesday:

1) Age: 17 out of 49 are over 50. As a percentage = 34.7%. Sex: 32 out of 49 are female, so 17 are male. 17/49 = 34.7%

2) Weight being carried = 850 – 470 = 380 kg. The car weighs 470 kg and 80% of this is 376 kg. The car is therefore carrying 4 kg more than its safe limit.

3) The ratio adds up to 5 + 11 + 2 = 18. Divide £2.7 million by 18 to get the size of a single share = £150,000, then multiply by the ratio numbers. The first ex-wife gets £750,000, the second gets £1,650,000 and the third gets £300,000.

4) Divide 5 litres (5,000 ml) by 40 to get the size of one part. 5,000 ÷ 40 ml = 125 ml. Multiply this by 3 to get the oil quantity: 125 × 3 = 375 ml.

5) c. (Jake got 3 parts more than Sam. $9,000 ÷ 3 = $3,000. 5 + 3 + 2 + 6 = 16; 16 × $3,000 = $48,000.)

6) b. (To save working with fractions, assume that there are 60 boys and 50 girls. Increased by 10% and 20% gives you 60 + 6 and 50 + 10 = 66:60, which simplifies to 11:10.)

7) a. (The plane is covering 300 miles in an hour against a 100 mph headwind, so its airspeed must be 400 mph. With a back wind its speed will be 500 mph. 1,800 ÷ 500 = 18 ÷ 5 = 3.6 = 3 h 36 m.)

8) c. (Euros were worth less than dollars at the start and more at the end, so there must be a gain. You can divide 0.70 by 0.63 to get the dollar to euro rate, but the working is clearer if you convert to pounds, then to euros (and vice versa).)

Thursday:

1) b (add two numbers to make the next)

2) d (two series, each × 3)

3) c (steps are 1, 2, 5, 1, 3, 5, 1, 4, 5, etc.)

4) d (steps are 1, 2, 2, 3, 3, 3, 4, 4, 4, 4, 5, etc.)

5) a (steps are – 5, +1, -4, +2, -3, +3, -2, +4)

6) 25 (add the previous number to the one three back)

7) y, w (two sequences, one jumps 3 letters, the other jumps 2)

8) 65 (reverse the digits and add 1, so 46 leads to 64 + 1)

9) b (multiply the outside digits to get the middle value)

10) a (move the digits to a set pattern)

Friday:

Bar chart

1) *Wheeler Dealer*
2) December (11 + 15 + 21 + 24 = 71; next is July: 8 + 13 + 18 + 27 = 66)
3) July (27 – 13 = 14; next is April: 21 – 10 = 11)
4) 21% (Aug = 14, Sept = 17, increase = 3; 3/14 = 21%)
5) Lower (*Top Gears* sales are consistently below average and falling. The other September sales are close to or above the previous October.)

Line graph

6) c
7) b. (Reading approximate rates from the graph, 500 euros ÷ 1.3 = £385; £385 × 1.2 = 460 euros.)
8) b
9) d. (Using approximate rates: $1,000 ÷ 2 = £500 × 2.1 = Fr 1,050. Fr 1,050 ÷ 1.4 = $750 × 1.6 = $1,200.)
10) c. (Using approximate rates: change = 2.0 – 1.4 = 0.6. As a percentage, 0.6/2.0 = 30%. The difference was slightly less than 0.6, so the answer must be a little under 30%.)

Pie chart

11) d. (Round up 29% to 30%. 160 × 30% = 48)
12) b. (Small sales = 2 × 46 = 92. Total sales = 160 + 46 = 206. 92 ÷ 206 = 0.446 = 45%.)
13) a. (Small: Round up 47 to 50. 50 × £5 = £250
Medium: Round down to 50% of 160 = 80 × £8 = £640
Large: 20% of 160 = 32 × £11 = £352
£250 + £640 + £352 = £1,242)
14) c (352 ÷ 1,235 rounds to 350 ÷ 1,200 × 100% = 350/12 and 360/12 = 30%)

Data from tables

15) 1,371 (43% of 3,189)
16) 0.27%. (In 1999 females = 54% of 3,357 = 1,813; in 2009 females = 57% of 3,189 = 1,818. Increase = 5 = 5 ÷ 1,813 × 100 = 0.27%.)
17) 1,436. (In 2004 young + old = 20 + 36 = 56%. Therefore those of working age = 44%. 44% of 3,263 = 1,436.)
18) 1,192 (17% of working age = 17% of 1,436 = 244. 1,436 – 244 = 1,192 employed.)
19) Cannot be calculated. (The sex and age/employment breakdowns are separate.)

ALSO AVAILABLE IN THE 'IN A WEEK' SERIES

For information about other titles in the 'In A Week' series, please visit
www.teachyourself.co.uk